Soil Chemistry

Second Edition

Shivanand Tolanur MSc (Agri) PhD
Professor and Head
Department of Soil Science and
Agricultural Chemistry
College of Agriculture
University of Agricultural Sciences
Vijayapur, Karnataka (India)

CBS

CBS Publishers & Distributors Pvt Ltd

New Delhi • Bengaluru • Chennai • Kochi • Kolkata • Mumbai
Hyderabad • Jharkhand • Nagpur • Patna • Pune • Uttarakhand

Soil Chemistry Second Edition

ISBN: 978-93-87085-07-7

Second Edition: 2018

First Edition: 2006

Published by Satish Kumar Jain and Produced by Varun Jain for
CBS Publishers & Distributors Pvt Ltd
4819/XI Prahlad Street, 24 Ansari Road, Daryaganj, New Delhi 110 002, India.

Ph: 23289259, 23266861, 23266867 Fax: 011-23243014 Website: www.cbspd.com
e-mail: delhi@cbspd.com; cbspubs@airtelmail.in.

Corporate Office: 204 FIE, Industrial Area, Patparganj, Delhi 110 092, India
Ph: 4934 4934 Fax: 4934 4935 e-mail: publishing@cbspd.com; publicity@cbspd.com

Branches

- **Bengaluru:** Seema House 2975, 17th Cross, K.R. Road, Banasankari 2nd Stage, Bengaluru 560 070, Karnataka
  Ph: +91-80-26771678/79 Fax: +91-80-26771680 e-mail: bangalore@cbspd.com
- **Chennai:** No. 7, Subbaraya Street, Shenoy Nagar, Chennai 600 030, Tamil Nadu
  Ph: +91-44-26680620, 26681266 Fax: +91-44-42032115 e-mail: chennai@cbspd.com
- **Kochi:** Ashana House, 39/1904, AM Thomas Road, Valanjambalam, Ernakulam 682 016, Kochi, Kerala
  Ph: +91-484-4059061-65,67 Fax: +91-484-4059065 e-mail: kochi@cbspd.com
- **Kolkata:** No. 6/B, Ground Floor, Rameswar Shaw Road, Kolkata-700014 (West Bengal), India
  Ph: +91-33-2289-1126, 2289-1127, 2289-1128 e-mail: kolkata@cbspd.com
- **Mumbai:** 83-C, Dr E Moses Road, Worli, Mumbai-400018, Maharashtra
  Ph: +91-22-24902340/41 Fax: +91-22-24902342 e-mail: mumbai@cbspd.com

Representatives

- **Hyderabad** 0-9885175004
- **Jharkhand** 0-9811541605
- **Nagpur** 0-9021734563
- **Patna** 0-9334159340
- **Pune** 0-9623451994
- **Uttarakhand** 0-9716462459

Printed at India Binding House, Noida, UP, India

Affectionately Dedicated to

My beloved Father

Late Sri IRAPPA

Wife PREMA

and

Daughters SHRUTI, SUSHMA

Preface

This textbook "Soil Chemistry" is the outgrowth of twenty four years of my teaching, research and extension experience in soil science.

The book "Soil Chemistry" comprises of eight chapters and an equal emphasis for discussion of various items is being given for all chapters. This book has been written by following the Syllabaii in the undergraduate studies of soil science of different Agricultural Universities of India and also it is intended to soil scientists, and other who are concerned with the studies of soil.

Each chapter of this book is arranged sequentially to facilitate readers is easy understanding the subject matter. Chapter 5th and 8th are the new chapters chemistry of Submerged soil and soil and water pollution have been included in this book which might be helpful for the students of agriculture in recent years.

It is hoped that this textbook will meet the growing need of soil science of different Indian Agricultural Universities. Any suggestions for improvement will be highly appreciated.

Dr. Shivanand Tolanur

Contents

Chapter ONE

Soil Colloids

SOIL AND ITS COMPONENTS

Soil may be defined as

"A dynamic natural body on the surface of earth in which plants grow, composed of mineral and organic materials and living forms"

- Buckman and Brady

The soil on which plants grow, provides

- a store house of minerals
- water
- air

Volume Composition of soil

The volume composition of soil in optimum condition for the crop growth are as follows :

Volume Composition of Soil — Total space (100 percent)

- Solid space (50 percent)
  - 45 percent mineral matter
  - 5 percent Organic matter
- Pore space (50 percent)
  - 25 percent water
  - 25 percent air

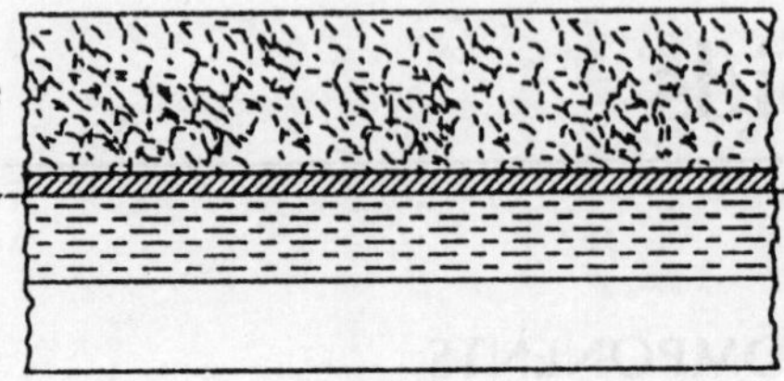

Fig. 1: Components of Soil

Soil (gives) helps

1. Mechanical support for crops
2. Store house of the plant food nutrients
3. Medium for plant growth
4. Seat of many chemical changes in presence of air, and water
5. Seat of micro-organisms

I. Mineral (inorganic) portion of the soil (1/2) of the soil volume is about

(50%) Soil solids

(25%) Soil Water

(25%) Soil Air

I. Mineral matter (45%)

Silica SiO_2

Aluminium & Iron oxides Al_2O_3 & $Fe_2O_3 \rightarrow 93\%$ — about 76% SiO_2 or more, 12% Al_2O_3, 5% Fe_2O_3

Oxides of Ca, Mg, K ----- 4% [1 + 1 + 1 + 1% MgO, CaO, K_2O, others]

Ti, Na, Trace of N, S, P. B, Mn, Zn]-- 3% [[all other constituents trace quantity] Cu, Mo, Cl, other elements]]

100

II. Organic matter (5%)

= Soil containing 20% or more organic matter that soil is called as organic soil

= Less than 20% organic matter referred as mineral soil

Mineral Soil (Inorganic soil)

◇ Less than 20% organic matter (generally less than 6% O.M.)

◇ Predominantly mineral or inorganic composition

◇ A mineral soil is a physical mixture of Inorganic particles, (sand, silt and clay) decaying organic matter (humus) and soil air, soil water

A Mineral Soil (Inorganic soil)

3 major substances or particles based on their size

1. Clay – Particle size colloidal (< 0.002 mm) dimension
2. Silt – Particle (face powder) size 0.02 to 0.002 mm
3. Sand – Particle that are gritty (2 to 0.02 mm)

Chemical ions in soil

◇ Cations in soils have +ve charge ex: Ca^{++}, Mg^{++}, K^+ etc.

◇ Anions in soils have –ve charge ex. CO^-_3, HCO^-_3, Cl^-, SO_4^- etc.

Soil Colloids

Consisting of clay and humified organic matter, inorganic (clay) colloids, organic colloids (humus). They are negatively charged. They attract (hold) positive charged particle called cation.

The colloidal state refers to a two phase system in which one material (or materials) in a very finely divided state is dispersed through a second Ex. Colloidal state milk and cheese, clouds and fog; starch, gelatin, rubber, proteins, blood, and soil etc.

- Solid in liquid : Clay in water (dispersion of clay in water)
- Liquid in gas : Fog or clouds – in atmosphere

The rapidly diffusing substances, mostly of crystalline nature in the solid state, are called crystalloids while starch and proteins are called colloids. But this distinction is not always true, because of crystalline substance like NaCl can be obtained in colloidal condition while albumin may be obtained by crystalline form. It is therefore be more appropriate to speak of a colloidal state of a substance than of a colloidal substance.

The size of the particles in true solution and in colloidal suspensions are = 0.2 –1 mμ and 1-200 mμ (micrometers) respectively. Colloidal particles – Float in a medium and – do not tend to settle. Therefore colloids are referred to as dispersed systems.

The substance in solution is called the dispersed phase while the medium in which the particles are dispersed is called the dispersion medium. Soil particles less than 0.001 mm size possess colloidal properties and are known as soil colloids.

Soil Colloids Importance in Soil Studies

1. Soil physical properties : The type of clay mineral present in soil determines to a large extent its physical behaviour, such as aggregation, volume expansion or swelling, plasticity, water retention and transmission characteristics or hydraulic conductivity, porosity, hygroscopicity and viscosity.

2. Soil Chemistry and nutrient availability : Physical chemistry of soil is actually chemistry of soil colloids (1) the clay (minerals) and (2) soil humus (organic) (1) clay minerals are mainly responsible for ion exchange reactions which take place between plant root and soil close to it or the rhizosphere (root).

3. Soil management : Important management practices such as irrigation, drainage and soil conservation which have to be adopted in a particular locality depend on the dominating clay mineral in the soil. So is the case of fertilizer application.

The highly swelling and cracking features of montmorillonite (Black soil) soils call for special types of soil management practices in comparision with those used for Kaolinitic and illitic soils (Red soil).

4. Soil pesticide interaction : Some of the pesticides applied as plant protection measures are adsorbed by soil particles (soil colloids).

5. Nutrient fixation : Illitic minerals are responsible for fixation of applied NH_4^+ and K^+ by soils.

6. Interaction with humic substances : Such interaction lead to the formation of clay humus complexes which are so important from the point of view of soil fertility.

7. Weathering and soil formation : Clay mineralogical make up of soil in conjunction with primary mineral assembleage indicate the stages and intensity of weathering and soil development.

8. Soil Classification : Clay minerals, because of their influence on soil properties are considered an important parameter for soil classification.

Nautre and Properties of Inorganic and Organic soil colloids

A colloid is any solid substance, whose particles are very small. The colloidal state refers to a two phase system in which one material in a very finely divided state is dispersed through second.

Examples are :

Solid in liquid : Clay in water (dispersion of clay in water)

Liquid in gas : Fog or clouds in atmosphere

The most important common property of inorganic and organic colloids are their extremely small size. They are too small to be seen with an ordinary light microscope. The Inorganic clay fraction of the soil contains particles less than 0.002 mm or 2 micrometers (μm) in diameter in size.

Nature of colloids : Soil colloids are of two kinds (1) Inorganic (minerals) and (2) Organic (humus). The two together form the colloidal complex of the soil. In almost all soils the inorganic colloids form a major portion of the colloidal complex.

Colloidal particles float in a medium and do not tend to settle. Some large colloidal particles may settle very slowly. Colloids are referred as the dispersed systems. The substance in solution is termed as the dispersed phase while the medium in which the particles are dispersed is called for dispersion medium. The commonest colloids are those containing minute solid or liquid particles suspended in liquid or gas medium.

(I) Inorganic colloids (clay)

The chemical analysis of clay indicates the presence of four main constituents silica alumina, iron, and the combined water. These make up from 90 to 98% of the colloidal clay. The colloidal matter of soil contains a higher proportion of the important plant nutrients such as Mg^{++}, Ca^{++}, Mg^{++} and K^{+}. The shape of the individual particle is plate or flake like

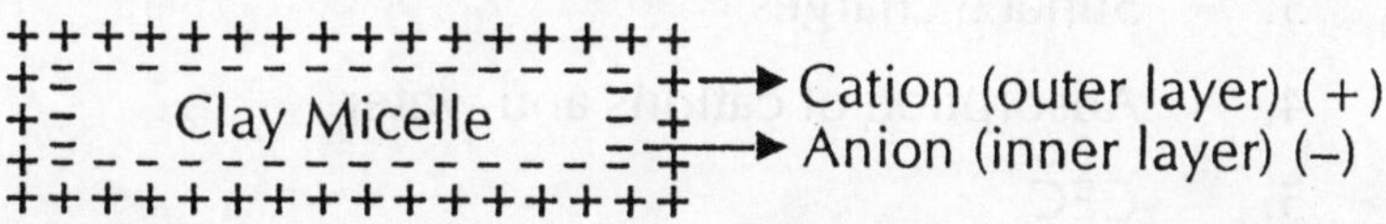

$^{+}$Cation
$^{-}$Anion

Clay colloids are : Negatively (-) charged (anions) and therefore attract a large number of + vely charged ions (cations). The minute clay colloids particles referred as micelles (microcells) ordinarily carry negative charge.

(II) Organic colloids (Humus)

Organic colloids are chiefly due to presence of humus in soil. Humus is the products of decomposition of plant and animal residues. Humus is amorphous dark brown to black nearly insoluble in water, but mostly soluble in dilute alkali (NaOH or KOH) solution.

Humus colloids are composed of carbon, (50%) Hydrogen, Oxygen(5%), and Nitrogen (30%) instead of silicon, aluminium and oxygen as in clay colloids. Lesser S,P and other elements are also present in humus.

Humus contains 30% nitrogen rich in proteins lignins and complex sugar (Polyuronides) varied organic substances cement soil aggregates together.

Organic soil colloids have higher adsorptive properties for water and cations (Ca^{++}, K^{+} etc) and higher cation exchange capacity than colloidal clay (Inorganic colloids).

General characteristics of soil colloids

1. Size
2. Surface area
3. Surface charges
4. Adsorption of cations and water
5. CEC

1. Size : Common property of inorganic and organic colloid is their extremely small size. They are too small to be seen with an ordinary light microscope.

Most are smaller than 2 micrometer (μm) (0.002 mm)

(1 micrometer (μm) or micron = 0.001 mm

1.0 mm = 0.04 inch

Thus, 25,400 ìm = 1.0 inch

2. Surface Area : Because of their small size, all soil colloids exposed a large external surface per unit mass.

The external surface area of 1 g of colloidal clay is at least 1000 times that of 1 g of coarse sand. The total surface area of soil colloids ranges from 10 m^2/g for clays with only external surfaces to more than 800 m^2/g for clays with extensive internal surfaces. The colloidal surface area in the upper 15 cm of a hectare of a clay soil could be as high as 700,000 km^2.

3. Surface charge : Soil colloidal surfaces, both external and internal, characteristically carry –ve and +ve charges. For most soil colloids, electronegative charges predominate, although some mineral colloids in very acid soils have a net electropositive charge.

The presence and intensity of the particle charges influence the attraction and repulsion of the particles towards each other thereby influencing both physical and chemical properties.

4. Adsorption of cations and water : The charges on soil colloids is the attraction of ions an opposite charge to the colloidal surfaces. Such attraction is of particular significance for negatively charged colloids. The colloidal particles, referred to as micelles (micro cells), attract hundred of thousands of positively charged ions, or cations, such as H^+, Al^{3+}, Ca^{2+}, Mg^{2+}. This gives rise to an ionic double layer (Fig 2).

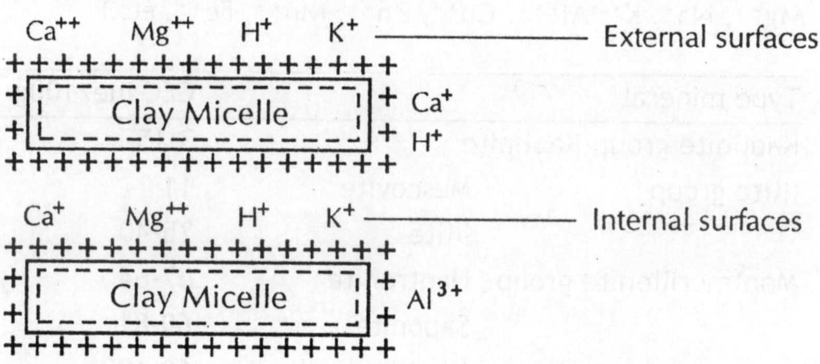

Fig. 2: Silicate clay particle (micelle) sheet like structure

In addition to the adsorbed cations a large number of water molecules are associated with soil colloidal particles. Some are attracted to the adsorbed cations each of which is hydrated. These water molecules play a critical role in determining both the physical and chemical properties of soil.

Specific surface area of some clay minerals

Clay (colloids) minerals	Range of surface area (m^2/g)		
	Internal	External	Total
Kaolinite	7-10	30-35	37-45
Illite	70-100	50-70	120-170
Montmorillonite	500-600	80-150	580-750
Chlorite	6-80	70-100	130-180
Vermiculite	700-800	80-100	780-900

5. Cation exchange capacity : Expressed in meq/100 g of the soil. Each clay colloids and organic colloids are characterized by their capacity to exchange cations called CEC (Cations. Ca^{++}, Mg^{++}, Na^{+}, K^{+}, Al^{+++}, Cu^{++}, Zn^{++}, Mn^{++}, Fe^{++}, etc.)

Type mineral		CEC me/100g
Kaolinite group-Kaolinite		3-15
Illite group	Muscovite	11
	Illite	25-40
Montmorillonite group	Nantronite	57-64
	Saponite	69-81
	Montomorrillonite	60-100
	Vermiculite	80-150
Humus (Organic colloids)		100-300

CEC of clay colloids and humus

Properties and Importance of Soil Colloids

1. Brownian movement
2. Flocculation

3. Electrical charge
4. Adsorption
5. Non-permeability
6. Cohesion and Adhesion
7. Swelling and Shrinkage
8. Plasticity

1. Brownian movement : Colloidal particles are found to be in continual motion. The oscillation is due to the collision of colloidal particles or molecules with those of the liquid in which they are suspended.

This movement is mainly responsible for the coagulation or flocculation of colloidal particles. When the particles in suspension colloid with each other and form a loose aggregate or floc.

2. Flocculation : The colloidal particles are coagulated by adding an oppositely charged ion. Formation of floc is known flocculation. If the cations are held close to the negatively charged particles, the negative charge would be neutralized and the colloidal particles flocculate and settle down Na^{+} (Sodium cations) are highly hydrated and are monovalent; they are not so closely bound with the negatively charged immobile particles. Thus, the particles continue to offer resistance to aggregation and do not flocculate Ca^{++} (Calcium cations) are divalent and are not easily displaced as sodium. Thus calcium ions are able to neutralize the negative charge more efficiently and the colloidal system tends to flocculate. In a similar manner trivalent ions like Aluminium (Al^{+++}) are still more efficient in flocculation of colloids. Thus Na clay produces deflocculation and Ca clay encourages aggregation.

The phenomenon of flocculation plays an important part

in the cultivation of soils. When clay particles are flocculated soil develops small clods of a crumby nature. Such a soil allows free movement of air and water. If the particles are deflocculated, the aggregates get dispersed, the soil gets waterlogged, the movement of air and water is impeded.

3. Electrical charge : Colloidal particles often have an electrical charge, some +ve and some –ve when clay colloids suspended in water, it carries a –ve electrical charge. Colloidal clay develops –ve electric charge due to dissociation of hydroxyl groups attached to silicon in silica sheets of the clay mineral leaves residual oxygen (O^-) carrying a negative charge.

4. Adsorption : Colloidal particles possess the power adsorbing gases, liquid and even solids from their suspension. The phenomenon of adsorption is confined to the surface of colloidal particles. Larger the surface area (area per unit weight) greater the adsorption for water, nutrients etc. For example, take a cube is equal to 6 square cm. When this cube is sub-divided into 8 cubes of edges 0.5 cm, the surface area exposed is 12 sq.cm.

Thus, the smaller the size of the particle the greater the surface area exposed by them. The adsorption of ions is governed by the type and nature of ion and type of colloidal particle. In the case of cations, the higher the valence of the ion the more strongly it is adsorbed. Exchange or replacement of cation would be difficult from colloidal particle. That is why divalent ions like calcium and magnesium (Ca^{++} and Mg^{++}) are held more strongly than monovalent ion sodium and potassium (Na^+ & K^+). Aluminium (Al^{+++}) a trivalent cation is most readily adsorbed. Hydrogen ions (H^+) behave as polyvalent ions so are adsorbed more strongly than even Ca^{++}

Adsorption of anions ($H_2PO^-_4$, HPO^-_4 etc) increases with the lowering or increasing of pH. The adsorption of phosphate ions is the lowest when the medium is neutral it increases when the pH either falls or rises, due to fixation by iron and aluminium by hydroxides in acid range and by calcium in alkaline range. Among clay minerals, Kaolinitic clay has a greater anion adsorbing capacity than montmorrillonitic or illitic clay.

The property of adsorption plays an important role in soil fertility. Due to this property soil is able to hold water and nutrients and keep them available to plant.

5. Non-permeability : Colloids, are unable to pass through a semi-permeable membrane. The membrane allows the passage of water and of the dissolved substance through its pores, but retains the colloidal particle.

6. Cohesion and Adhesion : Unlike sand, clay particles possess the properties of cohesion and adhesion. While forming aggregates, the colloidal clay particles unite with each other by virtue of the property of cohesion. Clay particles envelop sand particles under the force of adhesion. The force of cohesion and adhesion are developed in the presence of water. When colloidal substances are wetted, Water first adheres to the particles and then brings about cohesion between two or more adjacent colloidal particles. Soil when dried, the particles remain united because of the force of molecular cohesion. These two forces help in the retention of water in the soil and thus used by plants and micro-organism.

7. Swelling and Shrinkage : A soil colloid when brought in contact with water they imbibe a certain quantity of water and swell and increase in volume. Some clays such as smectite or montmorillonite swell when wet and shrink when dry. After a prolonged dry spell, soils high in smectites (eg. Vertisols) often are criss-crossed by wide deep cracks.

These properties of soils are largely responsible for the development and stability of soil structure.

8. Plasticity : Soil colloidal particles may present in gel condition possess the property of plasticity. Due to this property clay colloids can be moulded in any shape.

Soils containing >15% clay exhibit plasticity. This property is probably due to the plate like nature of the clay particles and the combined lubricating and binding influence of the adsorbed water. The particles of the plastic soils easily slide over each other, much like panes of glass with films of water between them.

Plasticity exhibited over a range moist to wet soil levels referred to as plasticity limits. At the lower of these levels termed plastic limit of the soil. The plastic limit is the lowest moisture content at which a soil cannot be deformed without cracking soils. Soils should not be tilled at moisture contents above the plastic limit.

The upper plastic limit or liquid limit is the moisture content at which soil ceases to be plastic, becomes semi-fluid (like softened butter) and tends to flow much like liquid. While these limits have only modest use for agricultural purpose.

Plasticity is of practical importance because of its influence on tillage operations. Thus, the cultivation of a fine textured soil when it is too wet will result in a "puddled" condition detrimental to suitable aeration and drainage with clayey soils, especially those of the smectite type, plasticity presents a significant problem. Stable granular structure is often difficult to establish and maintain in clayey soils high in smeetites.

Conclusion : Colloidal materials controls most of the chemical and physical properties of the soils. The specific types of inorganic and organic colloids in soils vary greatly in their capacities to adsorb and exchange nutrients and in their ability to be physically manipulated for both agricultural and non-agricultural purposes.

Type of Soil Colloids : Except Sand fraction, all soils contain particles of colloidal size. Soil colloids are chemically active portion of the soil.

Inorganic Colloids — Iron and aluminum hydrousoxide (occurs in tropical and subtropical region soils) Silicate clays (dominant in temperatre regions).

Organic Colloids — Humus (dominant in temperatre region soils).

The individual crystalline particle is called "micelle" or "micro cell". Colloidal clay particles are considered as ocupying an intermediate position being neither truely lyophilic (liquid loving) or hydrophillic (water loving) nor lyophobic (liquid hating) or hydrophobic (water repleelent). Clay colloids posses the properties of hydrophilic because of its hydration. Its sensitivity to electrolytes is a hydrophobic nature.

Lyophobic Colloids — The surface tension and viscosity of this colloidal system is similar to that of the dispersion medium. Small quantities of electrolytes cause precipitation. The particles are easily detected in the ultramicroscope and they migrate in one direction in an electronic field.

Lyophilic Colloids — The surface tension is usually lower and the viscosity is much higher than that of the dispersion medium. Large quantities of hte electrolytes cause prelipitation. The particles cannot be easily detected in Ultra microscope and they migrate in different directions or not at all in an electric field.

Among all kinds of soil colloids, the silitcate clays are the most dominant and important in the field of agriculture.

SILICATE CLAYS

Their Structure and Properties

The minerals occurring in rocks such as granite, basalt, gneiss and schist are termed Primary minerals. In the soil these (primary) minerals are decomposed to form secondary minerals, particularly clay minerals, generally minerals can be divided into non-silicates and silicates. The non-silicates include oxides, oxyhydroxides, sulphates, chlorides, carbonates and phosphates. Most have relatively simple structure but they vary widely in their solubility and resistance to weathering.

SILICATES

The silicate minerals have very complex structures in which the fundamental unit is the silicon-oxygen tetrahedron. This is composed of a central silicon ion surrounded by four closely packed and equally spaced oxygen ions – fourfold coordination. The whole forms a pyramidal structure, the base of which is composed of three oxygen ions with the forthforming the apex (Fig. 3).

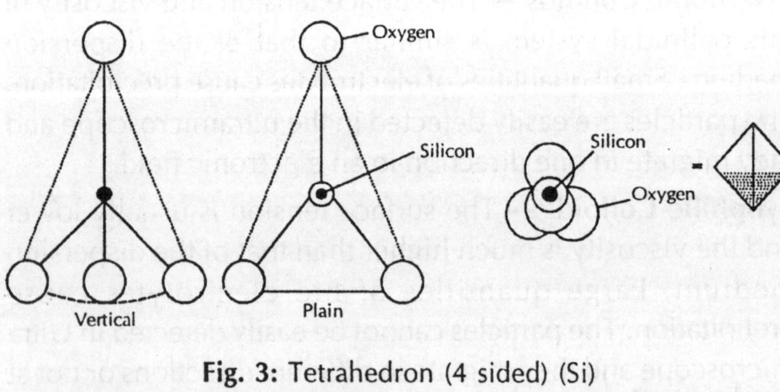

Fig. 3: Tetrahedron (4 sided) (Si)

A Silicon-oxygen tetrahedron

The 4 positive charges of Si^{4+} are balanced by four negative charges from the four oxygen ions. O^{2-} one from each ion, thus each discrete tetrahedron has 4 negative charges. An important variation in the tetrahedral structure is the substitution of Al for Si. This is known as isomorphous substitution (isomorphous replacement) which causes an imbalance in the charges within the structure, satisfied by cations such as Na, K, Ca, Mg.

Alumina-Magnesia octahedral

The 6 oxygens needed to fit snugly (compact and coordinate) around the larger Al^{3+} forms an 8 sided octahedron, and the Al^{3+} is in octahedral co-ordination (Fig. 4).

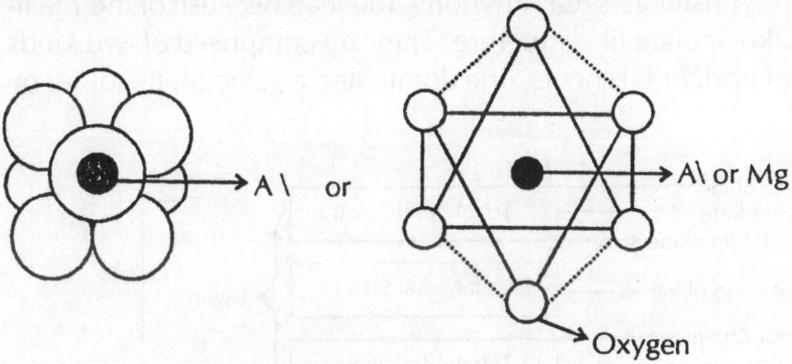

Fig. 4: 8 sided Octahedron (Al or Mg)

The tetrahedral and Octahedral sheets are the fundamental structural units of silicate clays. They inturn, are bound together within the crystals shared by oxygen atoms into different layers. The specific nature and combination of sheets in these layers vary from one type clay to another and largely control the physical and chemical properties of each clay.

Silicates are divided broadly into frame with silicates, chain silicates, ortho and ring silicates and sheet silicates.

Mineralogical organization of silicate clays

On the basis of the number and arrangement of tetrahedral (silica) and Octahedral (alumina-magnesia) sheets contained in the crystal units or layers, silicate clays are classified into three different groups (a) 1:1 type minerals– One tetrahedral (Si) to one octahedral (Al) sheet (b) 2:1 type minerals – two tetrahedral (Si) to one octahedral (Al) sheet) and (c) 2:1:1 type minerals – two tetrahedral (Si) to One octahedral (Al) to one octahedral (Al)

The most important silicate clays are known as phyllosilicates (Gr. Phyllon – the leaf) because of their leaf-like or plate like structure. They are comprised of two kinds of horizontal sheets, one dominated by silicon, the other by

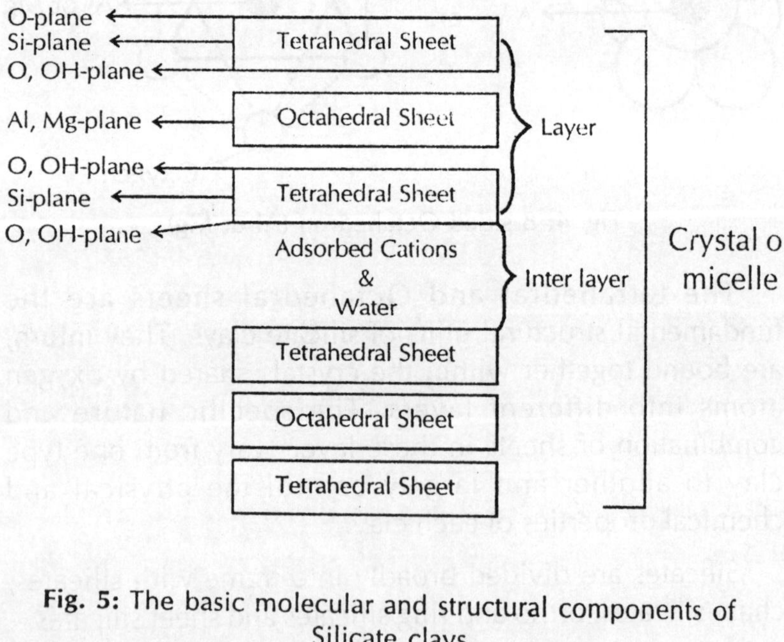

Fig. 5: The basic molecular and structural components of Silicate clays

aluminium and/or magnesium (Silica tetrahedral, Alumina octahedra). An Aluminium dominated sheet is known as dioctahedral sheet (2 Al ions) whereas one dominated by magnesium (3 Mg ions) is called as trioctahedral sheet fig.5.

(a) 1:1 Type minerals (Kaolin group) [Fig. 6]

- The thickness of the unit cell is 7.2A°
- Structural formula $Si_4 Al_4 O_{10} (OH)_8$

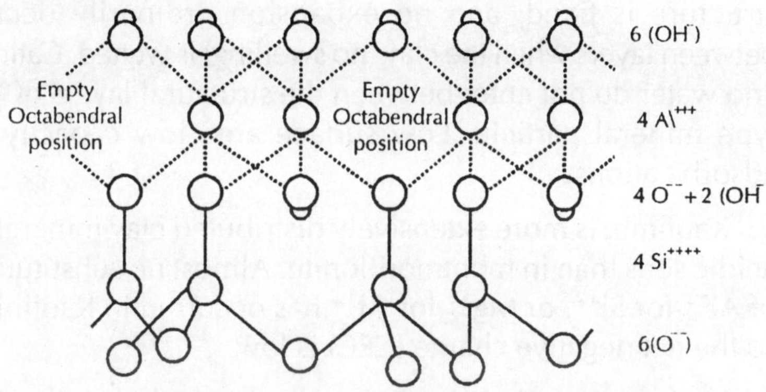

1:1 Type minerals (Kaolin group)

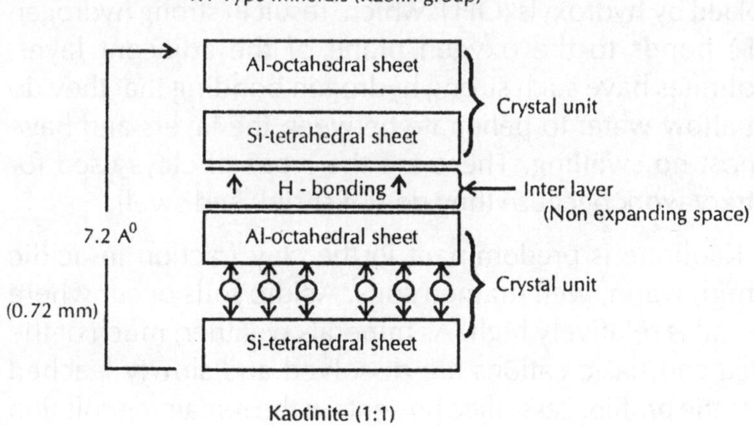

Kaotinite (1:1)

Fig. 6: Structure of Kaolinite ($Si_4Al_4O10\ (OH)_8$)

- ◇ 1 tetrahedral (Silica) : 1 octahedral sheet (Al)
- ◇ Member of Kaolin group – Halloysite, nacrite and dickite

The tetrahedral and Octahedral sheets in a layer of Kaolinite crystal are held together tightly by oxygen atoms, which are mutually shared by the silicon and aluminium cations in their respective sheets. These layers, in turn, are held together by Hydrogen bonding consequently, the structure is fixed, and no expansion ordinarily occurs between layers when the clay (no swelling) is wetted. Cations and water do not enter between the structural layers of 1:1 type mineral particle. Low surface area low capacity to adsorb cations.

Kaolinite is more extensively distributed clay mineral in acidic soils than in montmorillonite. Almost no substitution of Al^{3+} for Si^{4+}, or Mg^{2+} for Al^{3+} has occurred in Kaolinite, so the net negative charge (CEC) is low.

However, each layer has one plane of oxygen (O^{2-}) replaed by hydroxyls (OH^-), which, result in strong hydrogen (-H-) bonds to the oxygen plane of the adjacent layer. Kaolinites have such strong hydrogen bonding that they do not allow water to penetrate between the layers and have almost no swelling. These are the types of clays used for pottery work because they do not shrink and-swell.

Kaolinite is predominant in the clay fraction in acidic humid, warm, well drained soils. Acidic soils occur where rainfall is relatively high. As minerals weather, much of the silica and basic cations are dissolved and slowly leached from the profile. Less silica presents in the remaining solution means the proportion of alumina will be high enough to result in Kaolinite formation rather than the higher silica-containing montmorillonite. Kaolinite exhibits very little plasticity cohesion, shrinkage and swelling. Besides Kaolinite, there are other minerals in this type namely halloysite, ananxite and dickite.

(b) 2:1 type minerals

The crystal units (layers) of these minerals are characterized by an octahedral sheet sandwitched between two tetrahedral sheets.

Three general groups have this basic crystal structure. Two of them (1) Smectite and (2) Vermiculite include expanding type minerals, while the third (3) micas (illite) is relatively non-expanding

Expanding minerals

(1) Smectite group is noted for interlayer expansion, which occurs by swelling when the minerals are wetted, the water entering the interlayer space and forcing the layers apart. Montmorillonite is the most prominent member of this group in soils, although beidellite, nontronite and saponite are also found.

The flake like crystals of smectites are composed of 2:1 type layers (fig. 7(i)).

These layers are loosely held together by very weak oxygen to oxygen and cation to oxygen linkages. Exchangeable cations and associated water molecules are attracted between layers (the interlayer space), causing expansion of the crystal (unit) lattice (layer).

The total surface area per unit mass (external and internal) of one smectite mineral (montmorillonite) is 700-800 m^2/g. Kaolinite is only 5-20 m^2/g.

Isomorphous substitution of Mg^{2+} for some of the Al^{3+}

$Si_8 (Al_3 Mg) O_{20} (OH)_4 nH_2O$

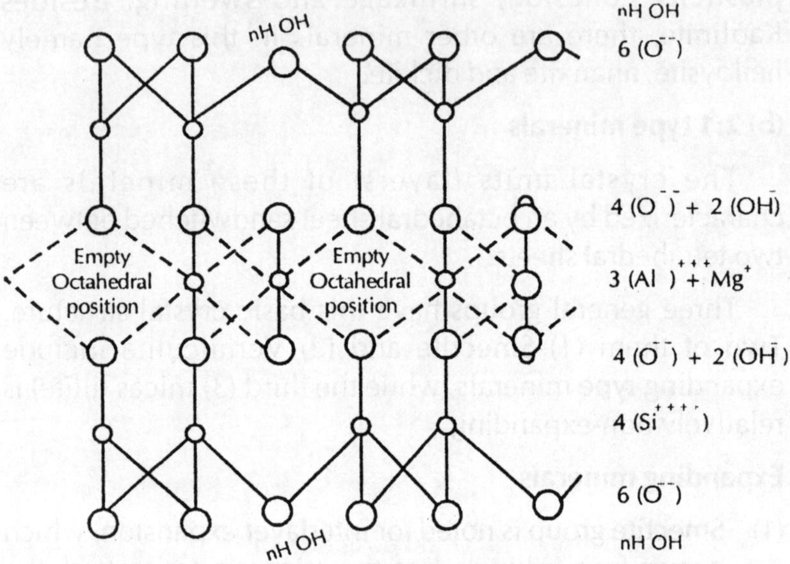

Fig. 7(i): Strcuture of Montmarillonite (2:1 type)

ions in the dioctahedral sheet accounts for most of the negative charge for smectites. The smectites show high CEC than Kalonite, smectite noted for their high plasticity and cohesion and their marked swelling when wet and shrinkage on drying, wide cracks commonly form as smectite dominated soils (e.g. Vertisols) are dried clods are very hard such soils difficult to till.

(2) Illite group : Fig 7(ii) (iii) One octahedral layer occurs between the two tetrahedral layers of Silica, one at the top and other at the bottom. The octahedral layer is joined with the tetrahedral layer when te top of the tetrahedral layer occupies the position after hydroxyl of the octahedral layer as shown in Fig. 7 (ii) & (iii). One-sixth of the silicons Si^{++++} has been replaced from

$K_2(Si_6\ Al_2)\ (Al_4)\ O_{20}(OH)_4$

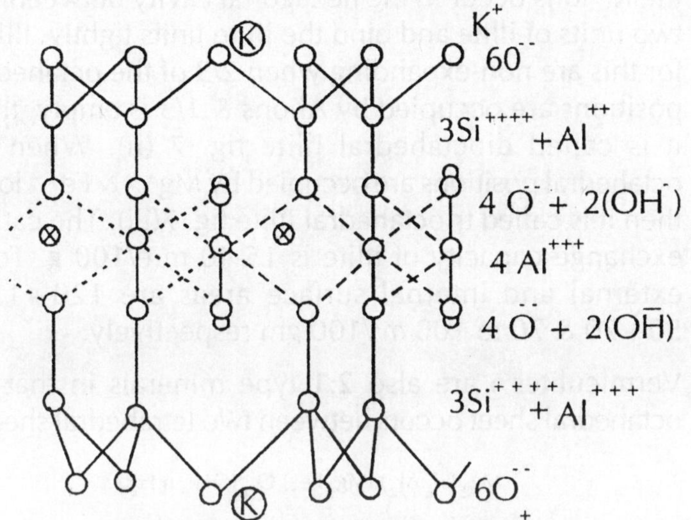

⊗ Empty octahedral position

Fig. 7(ii): Strcuture of dioctahedral illite

$K_2(Si_6\ Al_2)\ (Mg_4\ F3_3)\ O_{20}\ (OH)$

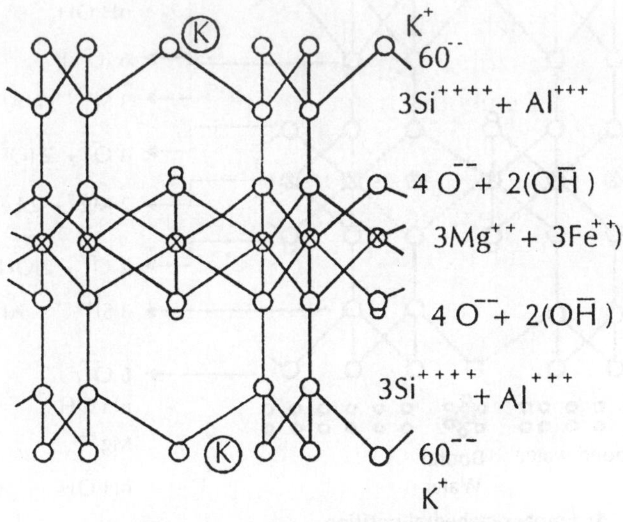

Fig. 7(iii): Strcuture of trioctahedral illite

tetrahedral positions by Aluminumions, Al^{++++}. Hence, the K^+ ions occur in the hexagonal cavity between the two units of illite and bind the illite units tightly. Illites for this are non-expanding when 2/3 of the octahedral positions are occupied by Alions & 1/3 is empty, then it is called dioctahedral illite fig. 7 (ii). When all octahedral positions are occupied by Mg^{++} & Fe^{++} ions, then it is called trioctahedral illite fig. 7(iii). The cation exchange capacity of illite is 15-40 m e/100 g. Total external and internal surface areas are 120+170, 50+70 & 70 to 100 m^2/100 gm respectively.

(3) Vermiculites : are also 2:1 type minerals in that an octahedral sheet occurs between two tetrahedral sheets.

$$Mg_2 (Si_6 Al_2) (Mg_2 Fe_2) O_{20} (OH)_4 \, n \, H_2O$$

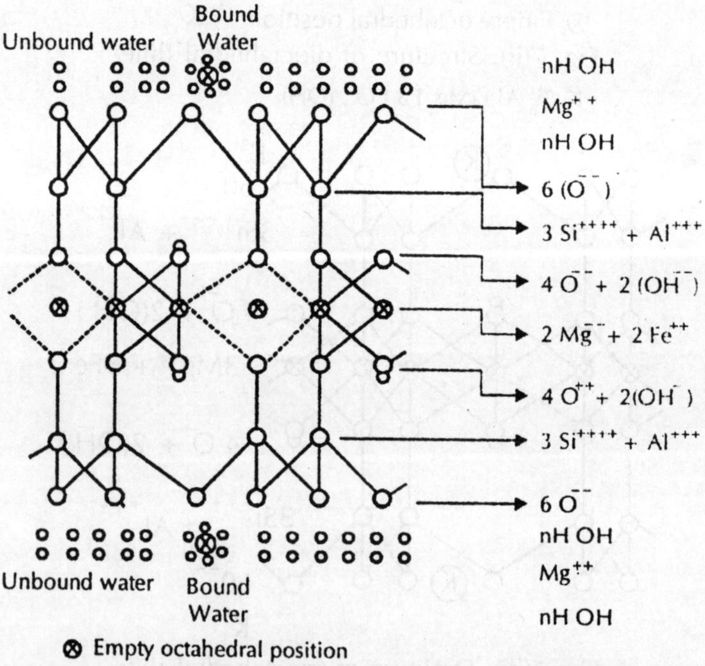

Fig. 8(i): Strcuture of diocthedral Vermiculite

In most soils vermiculites the octahedral sheet is aluminium dominated (dioctahedral) (Al) Mg dominated (trioctahedral) vermiculites are also common. Water molecules along with Mg and other ions are strongly adsorbed in the interlayer space of vermiculites (fig. 8(i) & (ii)).

The degree of swelling is considerably less for vermiculites than smectites (montmorillonite). Vermiculites have very high negative charge in the tetrahedral sheet. Therefore CEC is more that all of silicate clays.

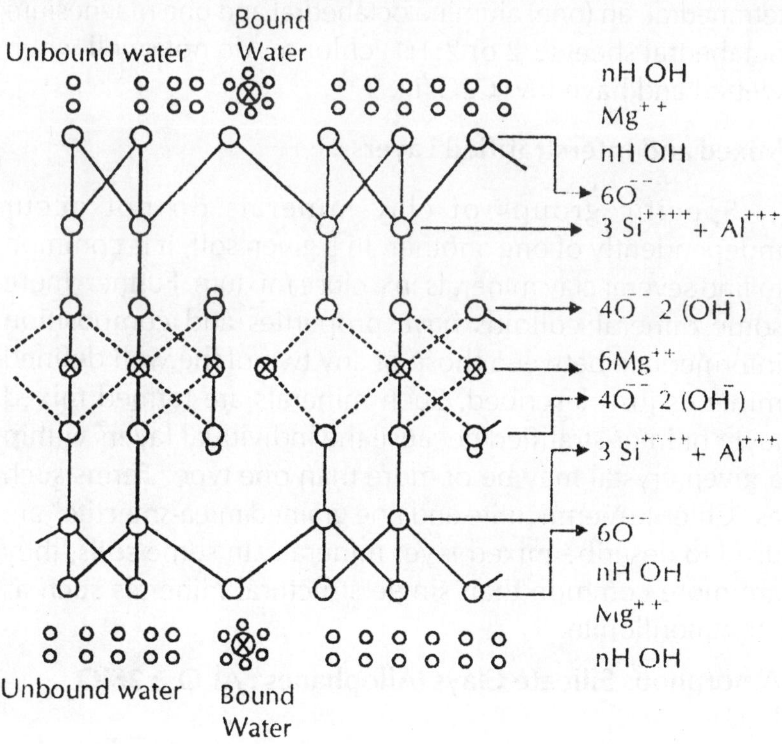

Fig. 8(ii): Strcuture of trioctahedral vermiculite

Non-expanding Minerals (2:1)

3. Micas are the type of minerals in this group. Muscovite and biotite are examples of unweathered micas often found in the sand and silt separates.

2:2 Type minerals

Chlorite clays are common in some soils. They are often called 2:1:1 type (or 2:2 type) clays because they are similar to the unit lattice of vermiculite except the hydrated Mg in chlorite is firmly bonded complete magnesium hydroxide octahedral sheet. Thus, a layer of chlorite has 2 silica tetrahedral, an (one) alumina octahedral and one magnesium octahedral sheet (2:2 or 2:1:1) chlorites do not swell when wetted and have low CEC fig. 9.

Mixed and Interstratified Layers

Specific groups of clay minerals do not occur independently of one another. In a given soil, it is common to find several clay minerals in a close mixture. Further- more some mineral colloids have properties and composition intermediate between those of any two of the well defined minerals just described. Such minerals are termed mixed layer or inter-stratified because the individual layers within a given crystal may be of more than one type. Terms such as "Chlorite-Vermiculite and fine grained mica-smectite" are used to describe mixed layer minerals. In some soils, they are more common than single structural minerals such as montmorillonite.

Amorphous Silicate Clays (Allophanes : $Al_2O_3 : 2SiO_2$ H_2O)

Amorphous silicate clays are mixture of silica and alumina

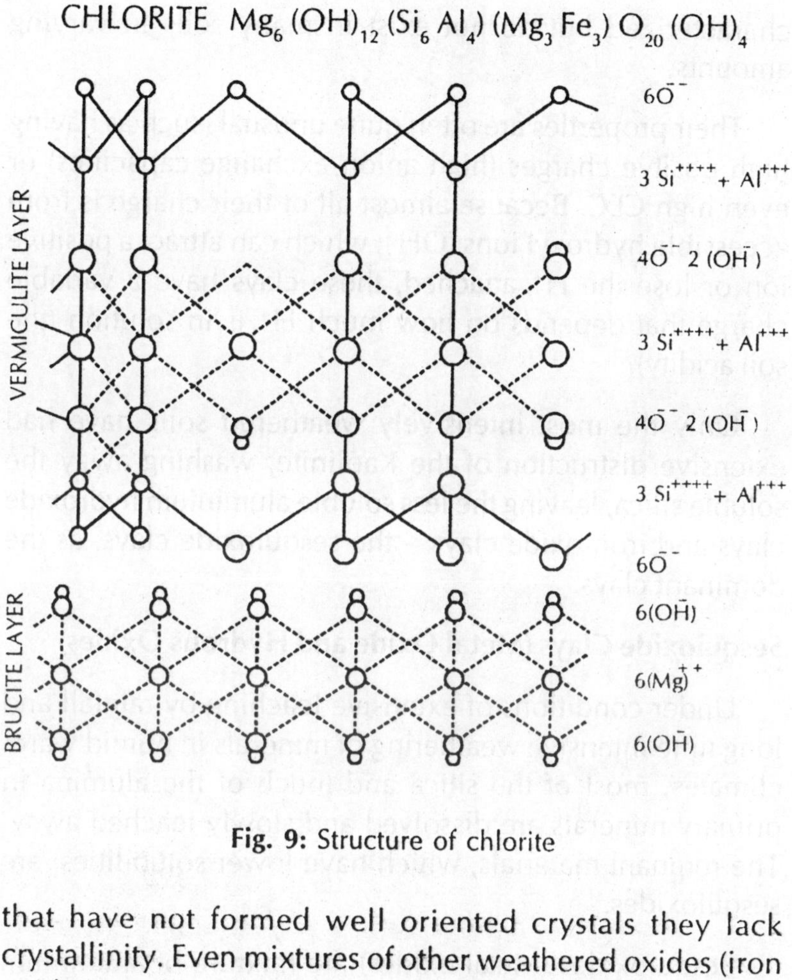

Fig. 9: Structure of chlorite

that have not formed well oriented crystals they lack crystallinity. Even mixtures of other weathered oxides (iron oxide) may be a part of the mixture. Typically, these clays occur where large amounts of weathered products existed but have not had the conditions or time for good crystal formation.

Amorphous clays are common in soils $Al_2O_3 : 2SiO_2\ H_2O$ forming from volcanic ash. Amorphous clays are not well characterized but do not exist in many soils in varying amounts.

Their properties are often quite unusual, such as having high positve charges (high anion exchange capacities) or even high CEC. Because almost all of their charge is from accessible hydroxyl ions (OH^-), which can attract a positive ion or lose the H^+ attached, these clays have a variable charge that depends on how much H^+ is in solution (the soil acidity).

Only the most intensively weathered soils have had extensive distruction of the Kaolinite, washing away the soluble silica, leaving the less soluble aluminium hydroxide clays and iron oxide clays – the sesquioxide clays, as the dominant clays.

Sesquioxide Clays (Metal Oxide and Hydrous Oxides)

Under conditions of extensive leaching by rainfall and long time intensive weathering of minerals in humid warm climates, most of the silica and much of the alumina in primary minerals are dissolved and slowly leached away. The remnant materials, which have lower solubilities, are sesquioxides.

Sesquioxides (Metal oxides) are mixture of aluminium hydroxide, $Al(OH)_3$ and ironoxides Fe_2O_3 or Iron hydroxides, $Fe(OH)_3$. (The Latin word Sesqui means one and one-half times). Sesquioxides refer to the clays of iron and aluminium because their formula can be written Al_2O_3 x H_2O and Fe_2O_3 x H_2O, one and one-half times more oxygen than Al or Fe. However TiO_2 and MnO_2 are also often included. These clays can grade from amorphous to crystalline.

Small amounts of these clays exist in many soils, with a relatively low extent of weathering. However, they will be the predominant clay and occur in large percentage only in soils formed in the humid, hot, well drained soils of tropical areas where intense weathering has occurred for perhaps hundreds of thousands of years.

These clays do not swell, are not sticky and have many differences from the silicate clays. Soils with 30-40% sesquioxide clays may absorb water almost as if they were fine sand. The sesquioxides coat large particles and form stable aggregates. The high percentage of iron and aluminium hydrous oxides furnishes an enormous surface for adsorption of phosphorus.

Organic Colloids (Humus)

Humus is a temporary intermediate product left after considerable decomposition of plant and animal remains – temporary because the organic substances remaining continue to decompose slowly. The humus is often referred to as an organic colloid and consists of various chains and loops of linked carbon atoms.

Humus is amorphous, dark brown to black, nearly insoluble in water, but mostly soluble in dilute alkali (NaOH or KOH) solutions. Humus contains 30% each of the nitrogen rich proteins, the slow to decompose lignins, and complex sugars (polyuronides). The polyuronides comprise much of the varied organic substances that cement soil aggregates together. Humus has about 50% carbon, some oxygen, 5% nitrogen, and less amounts of sulfur, phosphorus, and other elements. Humus also has a CEC on a dry weight basis many times greater than that of clay colloids. On a weight basis, a few per cent humus exerts much greater influence than does several times more per cent clay, especially for properties such as the cation exchange capacity.

Table 1 : Chief characteristics of the main groups of crystalline clay

Properties	Two layer clays 1:1 "Diphormic"	Three layer clays 2:1 (Triphormic)	
		Expanding lattice	Non-expanding lattice
1	2	3	4
Group members	Equidimen-tional Kaolinite nacrite dickite Elongate halloy site	Equidimentional Montmorillonite Sauconite beidellite Vermiculite Elongate Nontronite Saponite hectronite	Equipdementional Illite biotite Mascovite
Structure	Rigid lattice 1 tetra hedral layer for each octahedral layer	Lattice expanding and contracting with water content sheets made up of 1 octahedral layer with a tetra-hedralayer on both sides	Rigid lattice, sheets made up of 1 octahedral layer with a tetrahedral layer on both sides K bonds sheets together
Swelling and Shrinking with moisture changes	Very little	Much	Illite- little Micas - none
SiO_2/R_2O_3	1.9 - 2.1	4.0 - 6.0	2.1 - 3.2
Ionic substitution (Isomorphous substitution) in lattice	None	Al for Si, Mg, Fe, Mn, Ti for Al Li for Mg	Al for Si

1	2	3	4
Adsorptive capacity for in organic and organic, water, gases	Small	Large	Intermediate
Total -ve charge	5-10 meq/100g	80-120 meq/100g	About 250meq/100g
Cation(CEC) exchange capacity	5-10 meq/100g	80-120 meq/100g	Illite - 20-40 meq/100g Micas -2-5 meq/100g
Stability of H (Al) clay in a aqueous suspension	Low	High	Intermediate
Heat of wetting	1-2 Cal/g	10-20 Cal/g	Around 4 Cal/g
Specific surface	20-50 me^2/g	Around 750 m^2/g	75-125 m^2/g
Spacing Thickness with ethylene glycol	7.2A°	Montmorillonite 14A° Vermiculite 14A°	10A°
Temperature of first endo-thermic depression	600° C	100-200 °C	Illite - 50-100 °C Biotite - 100-400 °C Muscovite - 900 °C

The significance of clays in soils

The presence of clay in soils is of outstanding importance to their physical, chemical and biological behaviour. The great specific surface of clay makes it the active ingredient along with humus in the soil.

The cohesiveness, swell and shrink accounts for the important physical role of clay in the soil. Clays act as cement for sand and silt, binding these ingredients together.

Shrinking upon drying separates these soil masses, causing aggregated structure. The percentage of clay needed to cause this shrinking of the soil depends on the type of clay, its exchangeable cations, and the proportion of sand and silt in the soil. Silt without clay has very small pores that are only slowly penetrated by water. Silt is also highly erodible. The same is true of a mixture of sand and silt without clay.

Sand without clay has only negligible water retention power. It should be kept in mind that presence of a fair amount of organic matter would alter these situations. The most important chemical property of clays is their CEC, which to a large extent is the basis of plant nutrition. Some of the 3 layer clays especially montmorillonite and illite, have the ability to fix K ions in the interlayer spaces, sometimes with such an energy that the K is not released until the clay is dissolved. If there is a generous amount of humus in the soil K fixation is less severe 2 layer clays do not fix K.

All clays fix phosphate, but to very different degrees. The larger the specific surface and the higher the aluminium and iron content the greater is the fixation. For these reasons fixation increases in the order Kaolinite montmorillonite, allophane.

Clays also contribute to the nutrition of plants by the release of ions upon destruction of the clay crystals by weathering.

Origin of Charges on Soil Colloids

Adsorption and exchange phenomenon theory of crystal layer, theory of diffuse double layer

There are generally 2 types of charges i.e. one pH dependent and the other pH independent originate from exposed crystal surfaces (as a result of ionizable hydrogen ions) and isomorphous substitution respectively.

i) Exposed Crystal edges : The negative charge on the silicate clays develops due to unsatisfied valences at the broken edges of silica and aluminium sheet. Besides, the flat external surfaces of silicate clay minerals also serve as the source of negative charge. Ionizable hydrogen ions are hydrogen from hydroxyl ions on clay surfaces. The Al-OH or Si-OH portion of the clay ionizes the H and leaves a temporary negative charge on the oxygen (- Al – O^- or – Si – O^-). The machanism is shown below (Fig. 10) reaction are as follows :

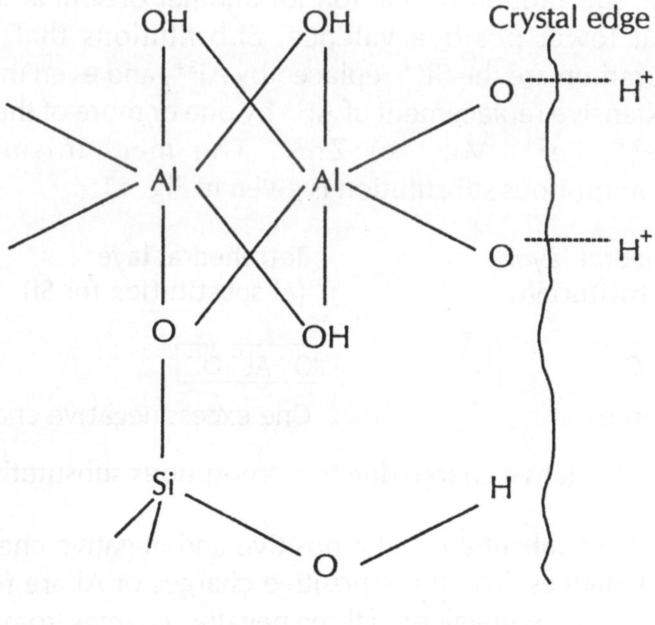

Fig. 10: Development of negative charge at the exposed crystal edges

$$- SiOH + OH \rightleftharpoons - SiO^- + H_2O$$

$$- AlOH + OH \rightleftharpoons - AlO^- + H_2O$$

Soil colloids	Soil Solution (no charge)	(negative (Soil Solution) charge Soil colloids)

The extent of ionized hydrogen depends on solution pH: more ionization occurs in more alkaline solutions. The magnitude of this pH dependent charge varies with the type of colloid. This type of charge is dominant for organic colloids.

ii) Isomorphous substitution : Isomorphous substitution is the substitution of one ion for another of similar size but lower positive valence. Substitutions that are common are the Si^{4+} replaced by Al^{3+}, and even more extensive replacement of Al^{3+} by one or more of these. Fe^{3+}, Fe^{2+}, Mg^{2+} or Zn^{2+}. The mechanism of isomorphous substitution is given in Fig. 11.

Tetrahedral layer (No substitution)	Tetrahedral layer (Al substitution for Si)
$O^{..}\ Si^{4+}\ O^{..}$	$O^{..}\ Al^{3+}\ O^{..}$
No charge	One excess negative charge

Fig 11: Negative charge due to isomorphous substitution

Without substitution, the positive and negative energy are in balances. The three positive charges of Al are fully satisfied by an equivalent of three negative charges from the surrounding oxygens or hydroxyls. There is no net negative or positive charge. However, when a Magnesium (Mg^{2+}) ion

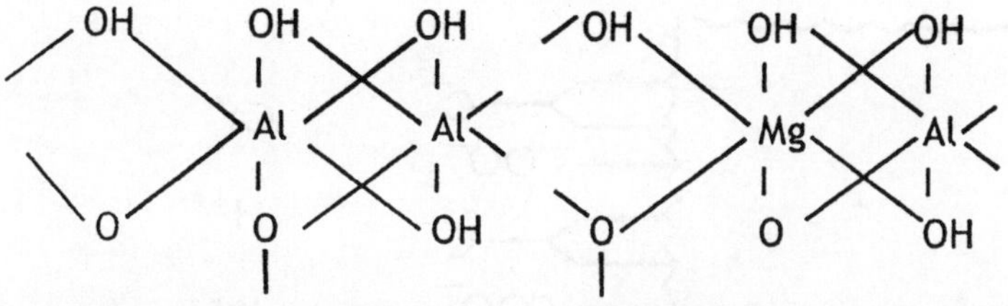

Without substitution With substitution of Mg^{++} for Al^{3+}

Fig. 12: Isomorphos Substitution

of about same diameter as an aluminium ion (Al^{3+}) replaces one of the aluminium by isomorphous substitution, an imbalance occurs. Consequently, the aluminium octahedral sheet assumes one negative charge for each Mg^{2+} for Al^{3+} substitution. In a similar way the negative charge develops due to isomorphous substitution in the silica tetrahedral sheet Fig. 12.

The charges arise from such isomorphous substitution are not dependent on pH and therefore, these charges are commonly referred to as permanent or pH independent charges.

iii) Anion exchange

Some clay minerals exhibit positive as well as negative charges. This will cause anion exchange between hydroxyl ions (OH) and anions like phosphate sulphate, chloride, etc.

Organic Soil Colloids

Humus is a temporary intermediate product left after considerable decomposition of plant and animal residues temporary because the organic substances remaining continue to decompose slowly. The humus is usually referred to as an organic colloid and consists of various chains and loops of linked carbon atoms Fig. 13.

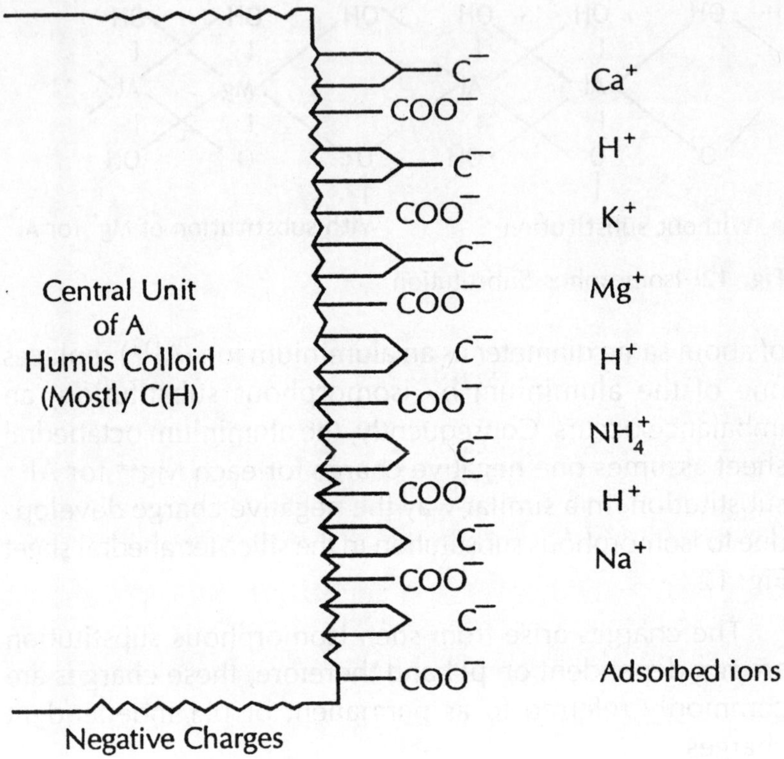

Fig. 13: Surface adsorption of cations by humus colloids

The phenolic hydroxyl groups (⊃O⁻) are attached to aromatic rings, the carboxylic groups (–COO⁻) are bonded to other carbon atoms in the central unit. Note the general similarity to the adsoption situation in silicate clays.

Humus micelle is composed basically of carbon, hydrogen and oxygen rather than of aluminium, silicon and oxygen in inorganic silicate clay minerals. The CEC of humus micelle is much higher as compared to montmorillonite (2:1 type) minerals. Humus is amorphous and the size of the individual micelle is variable. Humus is not stable as like that of clay and destroyed much more rapidly than clay. Humus has three components (1) Fulvic acid (2) Humic acid and (3) Humin.

The charge on humus colloids is pH dependent. Under strongly acid conditions hydrogen is rigidly bound and not easily replaceable by other cations. Humus colloid, therefore, exhibits a low negative charge.

Chapter TWO

Ion Exchange

ION EXCHANGE PHENOMENA

Ion exchange is defined as a reversible process by which cations and anions are exchanged between solid and liquid phases, and between solid phases it in close contact with each other. For exchange of cations and anions the term is used as cation and anion exchange respectively.

Adsorption is defined as a phenomenon by which an increase in concentration or an accumulation of an ion species on a solid occurs due to ion exchange or other reactions.

Desorption is a phenomenon by which the replacement or release of an adsorbed ion species occurs.

Chemical and physical processes are intimately related with ion exchange. Phenomenon which includes weathering of minerals, nutrient adsorption by plants, swelling and shrinkage of clay and leaching of electrolytes. Ion exchange particularly cation exchange, therefore, is considered as the most important process in soils.

Site for Ion Exchange

The ion exchange property of a soil is almost

entirely to the clay and silt fractions and the organic matter. So it is well known that the colloidal material of the soil with effective particle diameters of less than 20 microns (0.002 mm) being the most important site for causing the ion exchange phenomena. The soil particles are in amphoteric nature and so soil particles have the power to hold both cations and anions. It is also evident that the charge density and potential are higher on edges and corners and in furrows and cavities as compared to flat (plane) surfaces and consequently ion exchange phenomena are concentrated more in those sites.

Kind of Exchangeable Ions

The electric charge on the soil particles is neutralized by an equivalent amount of oppositely charged ions known as exchangeable counter ions, which are held to the surface mainly by coulomb forces and Van-der-waals forces.

In most of the soils exchangeable cations are Ca^{++}, Mg^{++}, H^+, K^+, Na^+ and NH^+_4 of which Ca^{2+} is the dominant exchangeable ion. In strong acid soils ($Al(OH^{++}_2)$) may act as a dominant exchangeable ions. Whereas in alkali soils sodium ion (Na^+) becomes dominant and the most common anions are SO^{2-}_4, Cl^-, HCO_3^-, etc.

Factors Affecting Ion Exchange

1. Nature and amount of clay.
2. Organic matter content mineralogical composition of the soil.

   Functional group carboxyl, phenolic, enolic etc with varying amounts.
3. Soil reaction and liming etc.
4. Irregular shape of the clay particles as well as non-

uniform distribution charges throughout the particles which changes the surface charge density of colloidal clays.

5. Clay minerals having 2:1 expanding lattice type generally have high ion exchange especially cation exchange than that of 1:1 type of clay minerals (low surface charge density).
6. The application of liming material in the acid soil decreases the rate of ion exchange reactions because of varying degree of ionization.
7. Ion exchange capacity – CEC of soils increases with the increase in soil pH (decreased activity of H^+ ions in soils) because part of the exchange position on the soil colloidal materials are pH dependent.

Cation and Anion Exchange

Cation exchange: The exchangeable cations of soils are the portions of the cations associated with soil solids (exchange sites) that are subject to interchange with cation in soil solution. The process of interchange of cations in soil solution with those in exchangeable form is known as cation exchange.

The exchange of cations in the soil takes place between

1. Cations in the soil solution and those on the exchange complexes of soil (clay crystal and humus).
2. Cations released by plant roots and those on the exchange complexes of soil.
3. Cation on the surface of either two clay crystals, two organic colloids (humus) or a clay crystal and humus particle.

Cation exchange reaction is reversible and considered as ion redistribution.

From the above three points it is clear that cation exchange takes place on the surface of clay and humus colloids as well as on the surface of plant roots.

Simple example : A soil contains high amount of adsorbed Ca on the clay micelle. Now if it is desired to replace all Ca from the exchange complex (clay micelle), the soil is leached with NaCl solution so that Ca from the exchange complex is constantly being replaced by Na until an equilibrium is attained and thereby the clay micelle would be saturated with Na^+. Cation exchange phenomena is chemically equivalent. The reaction is shown below :

$$\boxed{\text{Clay micelle}}\ \text{Ca} + 2\,\text{NaCl} \rightleftharpoons \boxed{\text{Clay Micelle}}\begin{matrix}\text{Na}\\ \text{Na}\end{matrix} + CaCl_2$$

The most important cations in soil are Ca^{++} and Mg^{++}, H^+, Na^+, K^+ and Al^{3+}

Power of Replacement

All the cations are not adsorbed with equal tenacity for a clay saturated with a particular cation. Again all the exchange sites do not possess the same strength of negative charge at the particle surface. Some cations that are held most tightly (with the greater strength of adsorption), have relatively small size and/or higher charge per unit size, and thus can approach closer or be held more strongly to the charged site than other ions. The attraction of a charge falls off with the square of the distance between the two charge centers. Hydrogen is the most tenaciously bound by soil colloids. The power of replacement is

$$H^+ = Al^{3+} > Ca^{2+} > Mg^{2+} > K^+ = NH_4 > Na^+$$

Cation Exchange Capacity (CEC)

CEC is the amount of exchangeable cations per unit weight of dry soil. CEC is measured in milliequivalents (me) of cations per 100 gms of soil (recently C mol (P^+) kg^{-1} soil). So it is the capacity of soil colloidal material in exchanging all its cations with the cations of the soil solution. It is one of the important chemical properties of soils and is usually intimately related to soil fertility and plant nutrition.

Milliequivalent concept

For the measurement of CEC, the term equivalent or more especially "Milliequivalent" is used because the number of negative charge sites in a given soil sample does not change but the weights of the cations that may be adsorbed to those sites at one time do change because they have different weights.

The term "milliequivalent" may be defined as one milligram of hydrogen or the amount of any other ion that will combine with or displace it. The milliequivalent weight of a substance is one thousandth of its atomic weight. Thus, if a clay has a cation exchange capacity of 1 milliequivalent (1 me/100g) it is capable of exchanging of 1 m g of hydrogen or its equivalent for every 100 g of clay. The weight of one hectare furrow slice (depth 0-15 cm) is 2.2×10^6 kilograms. Therefore, one hectare furrow slice (0-15 cm) soil is capable of exchanging 22 kilograms of hydrogen.

Calculation 0.1 kg soil can exchange $\frac{1}{10^6}$ kg of Hydrogen

1 kg soil can exchange $\frac{1}{0.1 \times 10^6}$ kg of Hydrogen

2.2×10^6 kg soil can exchange $\frac{2.2 \times 10^6}{0.1 \times 10^6}$ = 22 kg of Hydrogen

As for example, calcium, this element has an atomic weight of 40 compared to 1 for hydrogen. Each Ca^{2+} ion has two charges (divalent) and this is equivalent to two H^+ ions. Therefore, the amount of calcium required to displace 1 mg of hydrogen is 40/2 or 20 mg. So 1 me of Ca^{2+} weight 20 mg. If 100 g of a certain clay is capable of exchanging of a total of 250 mg of calcium (Ca^{++}), the CEC is 250/20 or 12.5 me/100g or 12.5 C mol (P^+) kg^{-1}

Application of Milliequivalent concepts to the practical field

This concept can be used for the calculation of the amount of lime required in the field. For instance, 1 m.e. of hydrogen can be replaced on the colloids by 1 m.e. of $CaCO_3$ (lime stone). Then calculate the amount of lime ($CaCO_3$) required in kg/hectare furrow slice to replace that 1 m.e of hydrogen.

We know, 1 m e of H^+ / 100 g = 1 m e of $CaCO_3$/100 g

1 mg of H+/100 g can replace $\frac{100}{2}$ or 50 mg of $CaCO_3$/100 g

In other words

50 mg of $CaCO_3$ /100 g will be required to replace 1 mg of hydrogen/100g. Since 1 me of hydrogen /100 g can be expressed 22 kilograms of hydrogen per hectare furrow slice.

So, 1 me of $CaCO_3$/100 g can be expressed as 22 x 50 or 1100 kilograms of $CaCO_3$ per hectare furrow slice.

Factors Affecting Cation Exchange Capacity

Many factors that affect ion exchange and also affect CEC of the soil namely (1) Texture of the soil (2) Organic matter content, (3) amount and kind of clay etc.

For ex: Soils with large amounts of clay and organic matter will have higher cation as well as anion exchange capacities than sandy soils low in organic matter. Also, soils with predominately 2:1 colloids will have higher exchange capacities than soils with predominately 1:1 mineral colloids. Soil reaction or pH of the soil also affect the cation exchange capacity of the soil. Generally with the increase in soil pH, the cation exchange capacity increases.

The exchangeable aluminium and other polymeric hydroxy aluminium also affect the cation exchange capacity of the soil by changing the soil reaction or pH.

Percentage of Base Saturation of Soils

The percentage of the total CEC satisfied with basic cations is termed as base saturation. It is defined as the extent to which the exchange complex of a soil is saturated with exchangeable cations other than hydrogen and aluminium and it is expressed as a percentage of the total cation – exchange capacity

$$\% \text{ BS} = \frac{S}{T} \times 100$$

where, BS = Base saturation

S = m. e of basic cation per 100 g soil

T = Total exchange capacity m e per 100 g soil.

All cations are included in the cation exchange capacity but in case of base saturation, all cation excluding hydrogen (H^+) and aluminium, are considered. Because these two groups of cations contibute H^+ ions in the soil solution. Adsorbed H^+ contributes directly to the H^+ ion concentration in the soil solution. Aluminium ions (Al^{3+}) also contribute H^+ ions to the soil solution through hydrolysis as follows :

$Al^{3+} + H_2O \rightleftarrows Al(OH)^{2+} + H^+$

$Al(OH)^{2+} + H_2O \rightleftarrows Al(OH)_2^+ + H^+$

$Al(OH)_2^+ + H_2O \rightleftarrows Al(OH)_3 + H^+$

Therefore an aluminium (Al^{3+}) ion is not considered as basic cations and other cations are called exchangeable bases. The degree of base saturation is an important property of soil which usually reflects the extent of leaching and weathering of the soil. It increases at the expense of the exchange acidity (acidity develops due to adsorbed H^+ and Al^{3+} on the colloidal surfaces). Example— How base saturation is calculated? Assuming that the following ion quantities were found in the ammonium acetate extract from the leaching of 20 g of soil : $\frac{0.02 \times 10^3}{20} = 1$ m.e

Calcium (Ca^{2+})	= 0.02 g = 0.02 x 10^3 mg	= 1 me
Mg^{2+}	= 0.006 g = 0.006 x 10^3 mg	= 0.5 me
Na^+	= 0.0115 g = 0.0115 x 10^3 mg	= 0.5 me
K^+	= 0.0195 = 0.0195 x 10^3 mg	= 0.5 me
	Total	**= 2.5 me**

So the milliequivalents of (Ca^+ atomic weight 40, equivalent weight 40/2 = 20) ion present in 20 g of soil is :

In a similar way, the milliequivalent of other ions present in 20 g of soil is being calculated and the total amount of those basic cations is 2.5 me per 20 g of soil or 12.5 me per 100g $\left(\frac{2.5 \times 100}{20} = 12.5 \text{ m.e}\right)$ suppose the total cation exchange capacity of the soil was 15 me per 100g. So the percentage of base saturation will be calculated as follows :

$$\% BS = \frac{S}{T} \times 100 \qquad S = \text{basic cation per 100 g}$$

$$= \frac{12.5}{15} \times 100 \qquad T = \text{Total exchange capacity m.e per 100 g}$$

$$= 83.3$$

% BS = Percentage of the CEC that is satisfied by the base forming cations is termed as = % base saturation.

As a a general rule, the degree of base saturation of normal uncultivated soils is higher for arid than for humid region soils. But this is not always true, especially in humid regions, the degree of base saturation of soil formed from lime stones or basic igneous rocks is greater than that of soils formed from sand stones or acid igneous rocks. Base saturation is closely related to pH of soil and also to the level of soil fertility. The ease with which cations are adsorbed by plants is related to the degree of base saturation. For any given soil the availability of plant nutrients like Ca, Mg and potassium increases with the degree of base saturation. It is evident that soils with large amounts of organic colloids or 1:1 type clay colloids can supply the nutrient cations to plants much lower degree of BS than soils in 2:1 type clay colloids.

Relationship between per cent Base saturation and pH

There is a definite correlation exists between the percentage base saturation of a soil and its pH. The determination of soil pH is comparatively more easier than that of determining base saturation. If pH is determined the base saturation can than be estimated approximately. A low percentage base saturation means acidity, whereas a % BS approaching 100 will result in neutrality or alkalinity. As the base saturation is reduced due to leaching loss of calcium, the pH also decreased (Due to development of acidity) in a definite proportion. Within the range of soil pH 5-6, the ratio for humid temperate region mineral soils is roughly at 5% BS change for every 0.10 unit change in pH. Thus, if the % BS is 50% at pH 5.5, it should be 25 per cent and 75% at pH 5.0 and 6.0 respectively.

The percentage base saturation does provide numerical values of the amount of exchangeable hydrogen and aluminium ion species and this helps in predicting and amount of lime needed to neutralize the soil acidity.

Importance of CEC in Plant Nutrition

CEC is an important reaction in soil fertility, in correcting soil acidity and basicity in altering soil physical properties, plant roots, when they come in contact with colloidal particles, adsorb exchangeable cations directly by interchange or contact exchange between the root hairs and colloidal complex.

Soil acts as a large cation exchanger. Any water moving through the exchanger will lose many of its soluble cations to the exchanger and pick up those cations that have been replaced from the exchange sites. Cation nutrients such as K^+, ammonium (NH^+_4) and Ca^{++} will not move far in soil before they are adsorbed to the exchange sites. This reduces their leaching loss but also reduces the mobility in soil. When potential water pollutants, such as lead (Pb^{2+}) or cadmium (Cd^{2+}), are disposed of an soil, instead of leaching easily they are adsorbed to the cation exchange sites. They may finally form insoluble hydroxides, carbonates, or their forms, but cation exchange helps initially to make them immobile and "Cleanses" the percolating liquid and the environment.

The nature and content of exchangeable bases in a soil have an important bearing on its physical, chemical and biological properties.

In normal fertile soils the total exchangeable bases (Ca, Mg, K and Na) constitute about 80-90% of the total CEC. Out of all exchangeable bases, calcium is the dominant cation. The pH value of such soil varies from 6.5 – 7.5 which is most conducive for the availability of plant nutrients. When the proportion of exchangeable hydrogen is high it gives rise to acid soil. In such soil, clay is saturated with hydrogen cations and forms H-clay and it is called base unsaturated soil. It is less fertile.

Out of total exchangeable cations in soil, when exchangeable sodium occupies more than 15% it gives rise to soils of alkaline reactions. The pH value of such soil is usually greater than 8.5. The soil turns into a sodium saturated and forms Na-clay. This soil is less fertile.

Anion Exchange

Anion exchange capacity of a soil to adsorb or release anions under normal conditions.

Charges arises from two sources

1. The protonation or adding of hydrogen ions, hydroxyl groups on the edge of these minerals.
2. The exchange of the hydroxyl (-OH) groups for other anions eg. phosphate.

Factors affecting anion exchange

1. Soil pH : anions increase with decrease in soil pH
2. Valency of ions : Anion exchange increase with valence of anion but anion exchange decrease with increasing valence of cation
3. Type and amount of clay colloid

Anion exchange is much greater in 1:1 type clay minerals – containing hydrous oxides of Fe & Al than (montmorillonite) 2:1 type clay

4. Salt concentration : Anion exchange decrease with increasing salt concentration. Addition of salts suppresses the unequal ion distribution. In general the relative order of anion exchange is →

$O\bar{H} > H_2PO_4^- > SO_4^- > NO_3^-$

Importance of Anion Exchange

Among various anions, the exchangeable phosphate is of most importance since sulphates or nitrates are not retained in sufficient quantities. So the phenomenon of anion exchange is most important for the release of fixed phosphate in the soil and thereby increases its availability to plants. The anion exchange is carried out mainly by replacement of OH ions of the clay minerals. The reaction is as follows :

$$\boxed{\text{Clay micelle}}\ OH + H_2PO_4 \rightleftarrows \boxed{\text{Clay micelle}}\ H_2PO_4 + OH^-$$

Adsorbed

Such type of phosphate adsorption renders phosphate unavailable to plants. Since the anion exchange is reversible phenomenon apart of the adsorbed phosphate ions again become readily available when soils are dominated with OH ions i.e. means under high pH conditions (may be induced with liming).

The origin of OH ions is from silicate clay minerals and hydrous oxides of iron and aluminium that are present in the soil. Large quantities of soluble phosphate fixed by these minerals as follows :

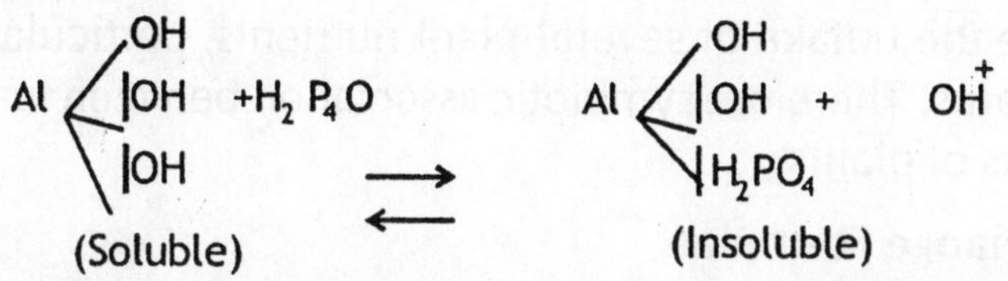

In this reaction takes place under slightly acidic soil conditions, exchange phenomenon is reversible, and soluble phosphate is further released when hydroxy phosphate compound(in soluble) comes in contact with – OH ions

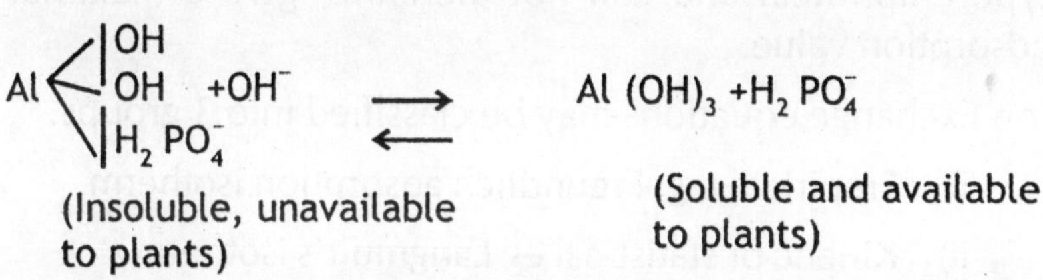

In this reaction occurs under moderate to strong acid conditions, the phosphate ions are irreversibly fixed and not available to plants.

Contact Exchange

Contact exchange is defined as the phenomenon by which the exchange takes place between ions held on the surface of solid phase particles (Solid-Solid Exchange) and that it does have to occur via the liquid phase. For ex: the exchange takes place between ions attached to the surface of root hairs (eg H^+ ions) and ions held on the surface of clays and organic matter in soils (because of the intimate contact that exists between roots and soil particles). Clay and plant roots both have CEC properties.

Importance of Contact Exchange

By contact exchange: It has been estimated that roots may grow to only about 3% of the available nutrients in soils. It enhance the uptake of several plant nutrients, particularly phosphorus. Therefore symbiotic association between fungi and roots of plants.

Ion Exchange Formulae

To describe ion exchange processes and predict on ion distributions, several formulae have been proposed. Among all formulae Freundlich's well known adsorption isotherm may be mentioned. The Freundlich formulae is of parabolic type emphirical and can not therefore, give a maximal adsorption value.

Ion Exchange equations may be classified into 3 groups.

i) Empirical e.g. Freundlich adsorption isotherm

ii) Kinetic or statistical eg Langmuir's isotherm

iii) Mass action e.g. Gapon Schofield's; Kerr; Hoffman and Griesse.

The principle of the law of mass action has been utilized to a great extent for deriving ion exchange formulae. Ion exchange phenomenon is not a real chemical reaction but merely a ion redistribution.

i) Empirical Formula : The Freundlich equation is one of the first used in soil studies and it states that —

$$m = AC^B$$

Where m = amount of adsorbed per unit weight of soil

C = ion concentration (ions to be allowed for adsorption) in soil solution

A & B are constants which vary from soil to soil

ii) Kinetic or statistical : The Langmuir equation is used to characterize ion adsorption particularly phosphorus in soils. The most common form of the Langmuir equation is

$$m = \frac{ABC}{1+Bc}$$

where, m = amount of ion adsorbed per unit weight of soil

c = ion concentration in soil solution to be considered

A = adsorption maximum

B = constant related to bonding energy

iii) Mass action

In homovalent (here monovalent – monovalent) exchange

$$RK^+ + Na^+ \rightleftarrows R\,Na + K^+$$

The equilibrium equation may be written as

$$\frac{[Na^+]_i\,(K^+)_o}{[K^+]_i\,(Na)_o} = K_{k\,Na}$$

For monovalent and divalent ions

$$RCa + 2K^+ \rightleftarrows 2\,RK + Ca^{2+}$$

At equilibrium

$$\frac{[K^+]^2_i\,(Ca^{2+})_o}{[Ca^{2+}]_i\,(K^+)_o} = K_{CaK}$$

Brackets denote concentratin; parentheses, activity; i, exchanger phase or inner solution; O, outer solution; K, the equilibrium constant which varies with the mole fractions of these two ions. Because of the difficulties encountered in the determination of activities of the ions present in the exchanger phase or inner solution, concentration has been used here.

Gapon Formula : All exchange equations are the same in the case of exchange of equal valence ions. This, however, is not the case when the ions are of unequal valences like monovalent and divalent ions

$$\frac{Nax}{Cax} \cdot \frac{(Ca)^{1/2}}{Na^{+}} = K$$

where, x is the exchange complex

K is the constant

This formula is used to characterize alkali soils and it also predict the alkali hazard of irrigation water. The Gapon constant K gives a measure of the relative tightness of binding of cations to clay.

Cation exchange is a reversible process in which one mole of cation charge (or equivalent) in solution replaces one mole of cation charge on the exchanger

ex : NH^{+}_{4} + Ca^{++} exchanger $\rightleftarrows$ NH_4 exchanger + ½ Ca^{2+}

NH^{+}_{4} + Na^{+} exchanger $\rightleftarrows$ NH_4 exchanger + Na^{+}

where NH^{+}_{4} and Ca^{++} are the molar activities of the ions in solution, and (Ca-exchanger) and NH_4 exchanger) represents the activities of the ions on the exchanger.

Cation exchange equation – gives the activity of the ions in the adsorbed state

NH^{+}_{4} + Ca exchanger $\rightleftarrows$ NH_4 exchanger ½ Ca^{++}

Assumed that NH^{+}_{4} and Ca^{++} are adsorbed in proportion to (a) the reduced ratio of their equilibrium activities in solution (b) the relative affinity of the exchanger for NH^{+}_{4} compared with Ca^{++}, as measured by a selectivity coefficient (K)

$$K_{NH_4\text{-}Ca} = \frac{NH_4\text{exchanger}}{\text{Ca exchanger}} = K_{NH_4\text{-}Ca}\,\frac{NH_4^{+}}{(Ca)^{1/2}}$$

The resultant equation is called the Gapon equation and has proved of wide applicability to cation exchange in soils. Provided that the monovalent cation (Na^{+} & K^{+}) amounts < 50% of the exchangeable Ca^{++} and Mg^{++} combined, reasonably constant values of the Gapon coefficients :

$K_{Na-Ca,Mg}$ and $K_{K\text{-}Ca\,Mg}$ are obtained for soils of similar clay mineral type from the equation.

$$\frac{\text{Na exchanger}}{\text{(Ca+Mg) exchanger}} = K_{Na\text{-}Ca+Mg}\,\frac{Na}{(Ca^{++} + Mg^{++})^{1/2}}$$

and

$$\frac{\text{K-exchanger}}{\text{(Ca+Mg) exchanger}} = K_{K\text{-}Ca\,\&\,Mg}\,\frac{K}{(Ca^{++} + Mg^{++})^{1/2}}$$

Schofield's Ratio Law

The Gapon equation indicates that if the proportions of

Na^+, & K^+ on the exchanger are unchanged, the ratio Na^+/K^+ in the equilibrium solution remains constant, and similarly, if the proportions of K^+ and Ca^{++} on the exchanger are constant, the ratio $\frac{K^{\cdot}}{(Ca^{\cdot})^{1/2}}$ is also constant. Schofield (1947) was the first to formulate such observations into a general hypothesis called the ratio law. It states that when cations in solution are in equilibrium with a large number of exchangeable cations, a change in the concentration of the solution will not disturb the equilibrium if the concentration of all the monovalent ions are changed in the one ratio, those of all the divalent ions in the square of that ratio and those of all the trivalent ions in the cube of Heat ratio.

"For concentration" one may read activity cation exchange between the double layer and soil solution may be described by the Gapon equation

$$\frac{\text{K-exchanger}}{\text{Ca-exchanger}} = K_{K\text{-}Ca} \frac{K}{(Ca^{++})^{1/2}}$$

Changes in solution activities that occur on leaching soils or in soils influenced by saline water, may be explained in terms of the ratio law which predicts that when cations in solution are in equilibrium with a large number of exchangeable cations, the activity ratio $a^+/(a^{n+})$ a constant over a limited range of salt concentration.

Elementary concepts of diffuse double layer

The exchangeable ions are surrounded by water molecules and may thus be considered as forming a solution which is often a micellar solution or inner solution. The solution containing free electrolytes called outer solution or inter-micellar solution. The ionic conditions on the outside surface of a clay particle or a pocket of clay particles dispersed in water or an electrolyte solution are controlled by the proportion of the exchangeable cations that disperse into the solution. A clay surface probably behaves as an effectively uncharged surface if the negative charge on the lattice is neutralized by monovalent cations that are tightly bound to the surface. But if the cations are hydrated a proportion tend to dissociate from the surface and will cause an electrical potential gradient to be set up near the surface. The system clay lattice exchangeable cations solution can be looked upon as forming a complex electrical double layer, known as the Helmholtz double layer, the inner layer being the surface of the lattice carrying the negative charge and the outer layer being composed of two parts a positive layer due to the cations bound to the lattice surface known as the fixed layer or stern layer, and a positive layer diffused in the solution close to the lattice surface known as the Gouy diffuse layer (Fig. 14).

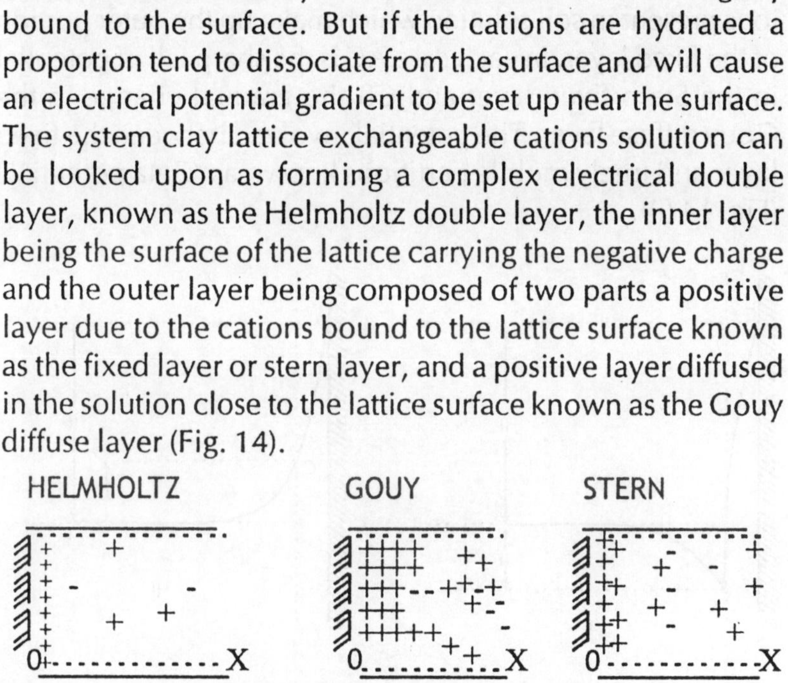

Fig. 14: Schematic representation of ion and potential distribution in the double layer according to the theories of Helmholtz Gouy & Stern.

The proportion of the cations that dissociate from the Stern into the Gouy layer depends partly on the concentration of salts dissolved in the water and partly on the tightness with which they are held to the surface. In the Stern fixed layer, there is a sharp fall of the potential vertical broken line, the extent of the fixed part of the double layer). If the potential of the solid is indicated by A and that of the bulk of the liquid by B, the fall of the potential in between may occur in two ways depending on the characteristics of ions present in soil solution which make up the outer portion of the fixed layer. In each case AC is the sharp fall of potential in the Stern fixed layer and CB the gradual change in the Gouy diffuse layer. The potential involved between the fixed layer and freely mobile partitions known as "Zeta potential" (Fig. 14 (i)).

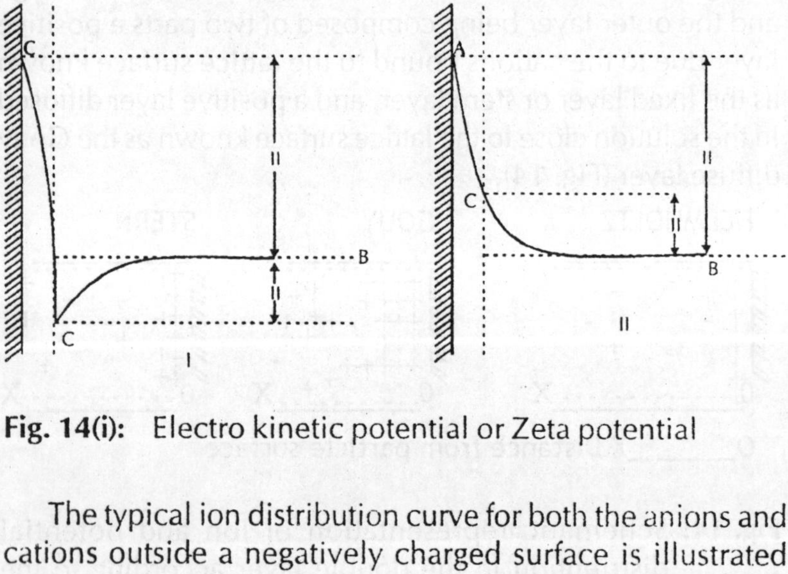

Fig. 14(i): Electro kinetic potential or Zeta potential

The typical ion distribution curve for both the anions and cations outside a negatively charged surface is illustrated (Fig. 14 (ii)) here. Assuming that Helmholtz layer has a uniform negative charge density and this assumption is reasonable when fairly large portion of the exchangeable cations will be dissociated into the solution and the double layer will thick, but it will become increasingly invalid as

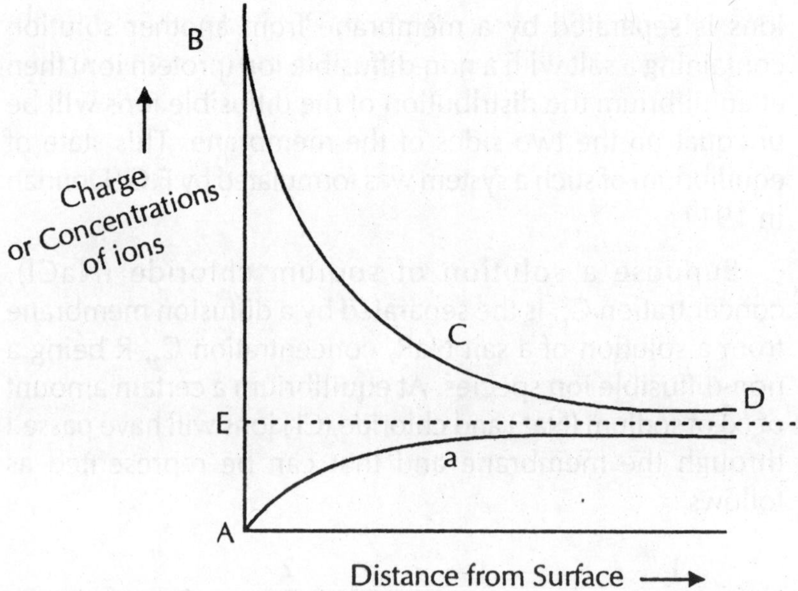

Fig. 14(ii): Ion distribution curve for anions and cations in a diffuse double layer AB: negative charge on the surface. BD: concentration of cations in the solution. AD : concentration of anions in the solution.

the proportion of cations dissociating from the surface drops, so the distance apart between the free negative charges on the surface becomes large. The charge density of the particle surface is equal to the sum of charges in the inner layer and in the outer layer. With the increase in concentration of ions in the solution, the stern fixed layer will charge and as a result ion swarm shrinks (thinner the double layer). On the other hand with the decreasing ionic concentration that means diluting the concentration of the solution ion swarm expands (thicker the double layer).

DONNAN MEMBRANE EQUILIBRIUM

If a solution of an electrolyte consisting of two diffusible ions is separated by a membrane from another solution containing a salt with a non-diffusible ion (protein ion) then at equilibrium the distribution of the diffusible ions will be unequal on the two sides of the membrane. This state of equilibrium of such a system was formulated by F.G. Donnan in 1911.

Suppose a solution of sodium chloride (NaCl), concentration C_1, is the separated by a diffusion membrane from a solution of a salt NaR, concentration C_2, R being a non-diffusible ion species. At equilibrium a certain amount of (x) of sodium (Na^+) and chloride (Cl^-) ions will have passed through the membrane and that can be represented as follows :

1		|	2		
Na^+	Cl^-	|	Na^+	R^-	Cl^-
$(C_1\text{-}x)$	$(C_1\text{-}x)$	|	$(C_2\text{+}x)$	C_2	x

|← - - - - - - - - - - - - - - - membrane

direction

flow of Na^+

and Cl^- ions

At equilibrium (total concentration of positive ions is equal to that of negative ions in each solution).

At equilibrium, the chemical potential (μ) of all ion present on both the solution must be equal

μ NaCl (1) = μ NaCl(2)

The chemical potential of an electrolyte may be taken as the sum of the potentials of its ions hence

$\mu^o Na + RT\ n\ ^a Na^+ (1) + \mu^o Cl^- + RT\ in\ ^a Cl^-$

$$= \mu^o Na^{\cdot} + RT\, n\, {}^a Na\,(2) + \mu^o Cl^- + RT\, in\, {}^a Cl^-$$

$$^a Na(1) \times {}^a Cl(1) = {}^a Na(2) \times {}^a Cl^-(2)$$

where 'a' denotes activities of ions

If these are dilute, the activities may be replaced by concentrations

$$^C Na^+ (1) \times {}^C Cl^-_{(1)} = {}^c Na(2) \times {}^C Cl^- (2)$$

$$(C_1\text{-}x)\,(C_1\text{-}x) = (C_2 + x)\, x$$

$$\text{therefore } \frac{X}{C_1} = \frac{C_1}{C_2 + 2C_1}$$

The fraction $\frac{X}{C_1}$ gives the proportion of (NaCl) sodium chloride initially present which has diffused through and it is found to be smaller when the concentration of non-diffusible ion is higher. Donnan membrane equilibrium belongs to the mass action equations.

A colloid particle, for example, a clay particle with its surroundings diffuse double layer may be looked upon as a micro-Donnan system, where the attractive electric forces between particle surface and counter ions act as a restraint, causing a non-uniform distribution of counter ions in the inner solution.

For a system with more than one diffusible electrolyte, the activity of each must be constant throughout the aqueous phase. For an aqueous system containing the ions K^+, H^+, Ca^{2+}, Cl^-, OH^- and SO^{2-}_4 the equilibrium is expressed by

$$\frac{(K^+)i}{(K^+)o} = \frac{(H^+)i}{(H^+)o} = \frac{(Ca^{2+})i}{(Ca^{2T})o} = \frac{(Cl^-)o}{(Cl)i} = \frac{(OH^-)o}{(OH^-)i} = \frac{(SO^{-2}_4)o}{(SO^-_4)_i}$$

Important Properties of Diffuse Double Layer

There are four important properties which are described here :

i) The thickness of the diffuse double layer decreases as the electrolyte concentration in the water increased at equal equivalent concentrations of electrolyte, divalent cations decrease the thickness of the double layer more strongly than that of monovalent cations, and trivalent more strongly than divalent. If clay particles suspended in water containing very little electrolyte, the double layer is thicker (wider) if the exchangeable ions are sodium (Na^+) and K^+ (potassium) than if they are (Ca^{2+}) calcium or magnesium (Mg^{2+}), and it is more thicker (wider) for sodium than for potassium. That is why soils containing more sodium show high pH resulting from more thicker diffuse double layer.

ii) The cation concentration gradient in this diffuse double layer sets up an electrical potential difference between the surface of the clay particle and the bulk of solution, that is, work must be done to transfer a cation from the surface of the particle to the bulk of the solution. Hence, this double layer hinders the free interchange of cation between these two regions. Hydrogen ions can not have the same activity throughout these regions, and the thicker the double layer, the smaller will be the concentration of the hydrogen ions that can dissociate from the clay surface to the hydrogen ion concentration in the bulk solution. The more diffuse double layer is compressed on to the clay surface, the more freely will the hydrogen ions be able to move from the clay surface into the solution, and the lower will be its apparent pH. The activity of the H^+ ions in the stern fixed layer may be much higher in some soils than that of outside layer and the activity appeared to be about a hundred fold higher than in the solution.

iii) The most direct way of estimating the proportion of cations that dissociate into Gouy diffuse layer should be by measuring the activities of the cations around the clay particles using a suitable membrane electrodes.

iv) The concentration of the anions in the solution is lower in the diffuse double layer than in the bulk of the solution. Hence, if a dilute sodium chloride solution is added to a dry sodium clay, and if the clay does not absorb any chloride (Cl^-) the concentration of chloride (Cl^-) in the bulk solution will rise because chloride will be partially expelled from the diffuse double layer.

This phenomenon is known as negative adsorption and it creates great difficulties when adsorption of anions by soils is being measured.

Chapter THREE

Soil Reaction and Buffering

Soil reaction is one of the most important chemical characteristics of the soil solution that influence many physical, chemical and biological properties of soil. Soil reaction is also the most important factor for evaluating its production potential for most crops. Soil reaction is an indication of the acidity or basicity of the soil and it has been classified into 3 groups (i) Acidity (ii) Alkalinity, and (iii) Neutrality.

METHODS OF EXPRESSING SOIL REACTION

Since soil reaction is the relationship between the concentration of H^+ and OH^- ions in the soil suspension which will determine whether the soil is acidic or alkaline or neutral reaction. Soil reaction can be measured by determining the pH. The term pH is from the French 'Pouvoir hydrogene' or 'Hydrogen power'. So pH may be defined as the logarithm of the reciprocal of the H^+ ion activity and it can be expressed in moles per litre

$$pH = \log \frac{1}{A_{H+}} \text{ or } -\log A_{H+}$$

where A_{H+} is the hydrogen ion activity in moles per litre.

A solution with an H^+ ion activity of 0.001 M will have a pH of 3.0, one with an H^+ ion activity of 0.0001 M, 4.0 and so on.

The concept of pH may be explained with reference to pure water, which is amphoteric and in which H^+ & OH^- ions are in equilibrium with undissociated water molecules ($H^+ = OH^-$).

In any solution in which water is the solvent, the product of the concentration of these H^+ and OH^- ions is approximately 10^{-14} at 25°C.

$a_{H+} \times a_{OH} \approx C_{H+} \times C_{OH-} = K_W = 10^{-14}$ gram ion per litre determined experimentally at 25°C or when pure water (HOH) ionizes to H^+ and OH^- (a neutral solution), both H^+ and OH^- are in concentration of 10^{-7} moles per litre (A mole is one molecular weight of ion or of the molecules

$$HOH \rightleftarrows H^+ + OH^-$$

Applying the law of mass action

$$\frac{[H^+][OH^-]}{[HOH]} = 1 \times 10^{-14}, (H^+) = (OH^-) = 1 \times 10^{-7}$$

The negative logarithm of $[H^+]$ is 7 or pH 7

or

In any solution in which water is the solvent the product concentration of these H^+ and OH^- ions is approximately

10^{-14} at 25°C

$$H_2O = H^+ + OH^-$$

$[H^+] \times [OH^-] = 10^{-14}$ if pH is 7.0, the H^- ion concentration of OH^- ions will be $\frac{10^{-14}}{10^{-7}}$ M per lit. At neutrality, the H- ion concentration is 0.0000001 M 10^{-7} or 10^{-7} M per lit of solution

$$pH = \log \frac{1}{10^{-7}}$$

$$= \frac{1}{0.0000001}$$

$$= \log 1000000$$

$$= 7$$

From pH 7 to 0 the soil is increasingly more acidic, from pH 7 to 14 the soil is increasingly more alkaline (basic). The H^+ ion concentration has ten fold change between each whole pH number. Thus a soil of pH 6 has 100 times (10^2) more H^+ ions in solution than a solution with a pH of 7.0

The relationship between H^+ ion concentration and pH. pH and poH are presented

H^+	pH
1×10^{0}	0
1×10^{-1}	1
1×10^{-2}	2
1×10^{-3}	3
1×10^{-4}	4
1×10^{-5}	5
1×10^{-6}	6
1×10^{-7}	7 – Neutral
1×10^{-8}	8
1×10^{-9}	9
1×10^{-10}	10
1×10^{-11}	11

1×10^{-12}	12
1×10^{-13}	13
1×10^{-14}	14

Relationship between pH and pOH

pH	Acidity (Normality of OH^+)	Alkalinity (Normality of OH^-)	pOH
0	1.0	0.00000000000001	14
1	0.1		13
2	0.01		12
3	0.001		11
4	0.0001		10
5	0.00001		9
6	0.000001		8
7			7
8			6
9			5
10			4
11			3
12			2
13	0.0000000000001	0.1	1
14	0.00000000000001	1.0	0

The relationship between (1) H^+ and pH (2) and pH and OH ion concentration and pH

pH	H^+ ion concentration	pH (H ion activity)	OH^- ion concentration	pOH
0	1×10^0	0	1.0	14
1	1×10^{-1}	1	0.1	13
2	1×10^{-2}	2	0.01	12
3	1×10^{-3}	3	0.001	11
4	1×10^{-4}	4	0.0001	10
5	1×10^{-5}	5	0.00001	9
6	1×10^{-6}	6	0.000001	8
7	1×10^{-7}	7	0.0000001	7
8	1×10^{-8}	8	0.00000001	6
9	1×10^{-9}	9	0.000000001	5
10	1×10^{-10}	10	0.0000000001	4
11	1×10^{-11}	11	0.00000000001	3
12	1×10^{-12}	12	0.000000000001	2
13	1×10^{-13}	13	0.0000000000001	1
14	1×10^{-14}	14	0.00000000000001	0

$$pH = \log \frac{1}{AH_+} \quad H^+ \text{activity moles/lit}$$

$$= \log \ AH \quad H^+ \text{atomic wt. (1)}$$

[Molar solution which contains a gram mol. wt. of the substance dissolved in a lit of the solution]

Ranges in soil reaction

For mineral soils the extreme range in soil reaction or pH extends from 3.5 to 10.5. Some times in peat soils, soil reaction or pH may be low to 3.0 or less and some alkali soils the reaction or pH may be high as 11.0. The ranges in soil pH for most mineral soils are shown in fig. 15.

Factors Affecting Soil Reaction

There are various factors which influence the soil reaction viz. nature of soil colloids percentage base saturation and kind of adsorbed bases, soil solution, climate, soil management and oxidation reduction state of soil.

1) **Nature of soil colloids :** The colloidal particles of soil (clay and humus) influence soil reaction to a great extent. When hydrogen (H^+) ion forms the predominant adsorbed cations on soil colloids (clay and humus) the soil reaction becomes acidic.

2) **Percentage base saturation and kind of adsorbed bases:** Soils having low percentage base saturation show acidic reaction. Sodium saturated soils have much higher pH values than those dominated by calcium and magnesium.

3) **Soil solution :** The soil solution comes a number of salts dissolved in water. The cations of the salts intermingle or mix together with those of the diffuse double layer of the clay colloidal particles and increase the concentration.

In other words, the concentration of cations in the bulk of the solution is more or less the same as that near the particle surfaces.

For a base unsaturated soil, a large number of hydrogen

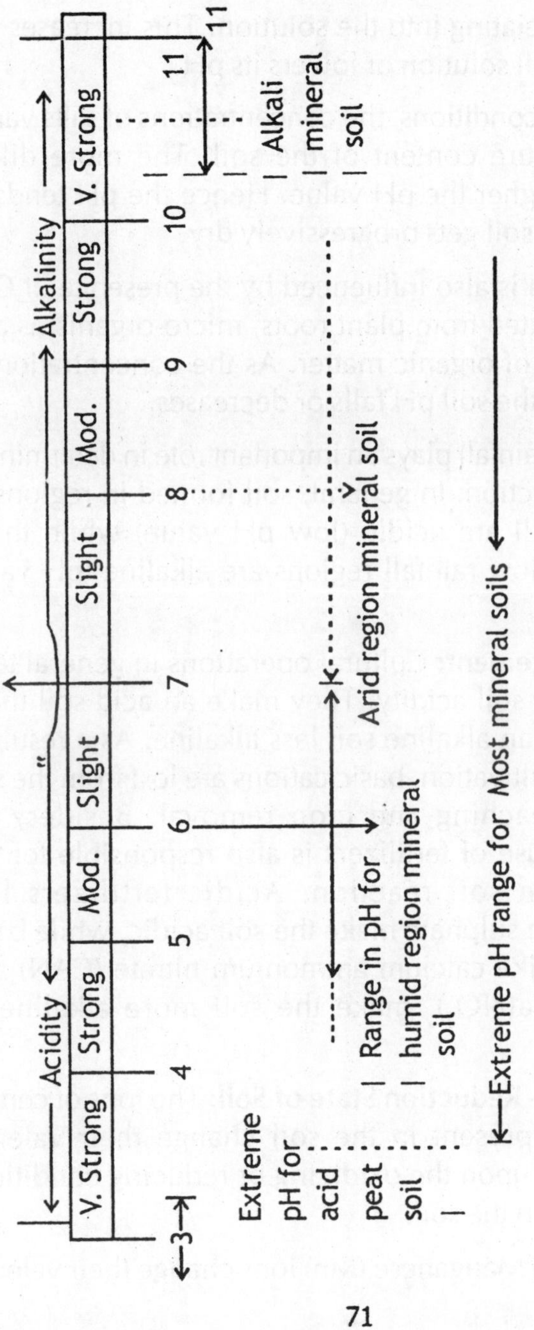

Fig. 15 : Extreme range in pH for most mineral soil and ranges comonly found in humid and aird region soils.

ions (H^+) dissociating into the solution. This increases the acidity of the soil solution or lowers its pH.

Under field conditions, the concentrations of salts varies with the moisture content of the soil. The more dilute solution, the higher the pH value. Hence the pH tends to decrease as the soil gets progressively dry.

Soil reaction is also influenced by the presence of CO_2 in soil air liberated from plant roots, micro-organisms and decomposition of organic matter. As the concentration of CO_2 increases, the soil pH falls or decreases.

4) **Climate :** Rainfall plays an important role in determining the soil reaction. In general, soil formed in regions of high rainfall are acidic (low pH value) while those formed in low rainfall regions are alkaline (pH value high).

5) **Soil management:** Cultural operations in general tend to increase soil acidity. They make an acid soil more acidic and an alkaline soil less alkaline. As a result of constant cultivation, basic cations are lost from the soil through leaching and crop removal. Besides, the continual use of fertilizers is also responsible for the changes in soil reaction. Acidic fertilizers like ammonium sulphate make the soil acidic, while basic fertilizers like calcium ammonium nitrate (CAN) and $NaNO_3$, $Ca(NO_3)_2$ make the soil more alkaline in reaction.

6) **Oxidation - Reduction State of Soil:** The ions of certain substance present in the soil change their valency depending upon the oxidizing or reducing conditions prevailing in the soil.

Iron (Fe) and Manganese (Mn) ions change their valency

with change in aeration. Under aerobic conditions these ions (Fe & Mn) are present in their oxidized state as their higher valence (Fe^{3+} and Mn^{4+}) whereas under anaerobic or reduced conditions their valencies are reduced to ferrous (Fe^{2+}) and manganous (Mn^{2+}) state. This brings about a change in soil reaction and the pH is increased. Under similar reduced conditions, sulphur is usually present as sulphide (S^{2-}). Upon aeration sulphur is converted to its oxidized sulphate (SO^{2-}_4) and is available to plant. This also increases the soil acidity and lowers the soil pH.

Influence of Soil Reaction on Nutrient Availability and Plant Growth

Plants can grow on soils with a wide range of soil reaction of pH as well as diversified types of soil conditions. The diversity in plants growing on soils with a varying range of pH is the best evidence. However, plants do have preferences for a specific range of soil reaction or soil pH. Soil reaction preferences for some crops grown in India are presented in Table 2.

Table 2: Preferences of Plants to Soil Reaction

Crops with Botanical names	Optimum soil reaction (pH range)
Cereals	6.0 – 7.5
Maize (Zea mays)	5.0 – 6.5
Millets	4.0 – 6.0
Rice (Oryza sativa L)	4.0 – 6.5
Sorghum (Sorghum vulgare)	6.0 – 7.5
Wheat (Triticum aestivum)	6.0 – 7.5
Barley (Hordeum vulgare L.)	6.0 – 7.5

Oats (Avena sativa L)	5.0 – 7.5
Rye (Secale cereale)	5.0 – 7.0
Legumes and Pulses	5.5 – 7.0
French bean (Phaseolus vulgaris L)	5.5 – 7.0
Berseem (Trifolium alexandrinum)	6.0 – 7.5
Groundnut (Arachis hypogaea L)	5.0 – 6.5
Soybean (Glycine max Merr)	5.5 – 7.0
Pea (Pisum sativum L)	5.5 – 7.0
Lentil, gram etc	5.5 – 7.0
Other field crops	
Sugarcane (Saccharum officinarum L.)	6.0 – 7.5
Cotton (Gossypium hirsutum)	5.0 – 5.5
Potato (Solanum tuberosum L.)	5.0 – 5.5
Sugarbeat (Beta vulgaris)	6.5 – 8.0
Tobacco (Nicotiana tabacum)	5.5 – 7.5
Cowpea (Vigna sinesis)	5.0 – 6.5
Sunflower (Helianthus annus)	6.0 – 7.5
Tea (Camellia sinensis L.)	4.0 – 6.0

Soil reaction and plant growth are interrelated as a result of the significant effect of soil reaction on soil environmental influence arising from increase or decrease in availability of nutrient element or from increasing concentration of certain plant nutrients to the toxic level eg. of Al^{3+} Fe^{2+} or Mn^{2+} in acid soils. The other environmental factor is soil physical condition which are rendered unfavourable under low or high pH.

Soil reaction (pH) is most important factor which governs availability of various essential as well as functional elements in soil by influencing the various soil properties like physical, chemical and biological etc.

The primary and secondary nutrients nitrogen, phosphorus, calcium and magnesium are as available or more available at a pH of 5.5 and 6.5 for organic and mineral soils than at any other pH. Molybdenum availability is significantly dependent on pH. In strongly acid soils it is quite unavailable. However, Molybdenum, boron and copper availabilities are also relatively high at a range of pH 5.5 to 6.5. The micronutrients like iron, manganese and zinc are less available at pH of 5.5 and 6.5 than at more acidic reactions.

Soil reaction and microbial acidity

It is evident that the activity of micro-organisms is influenced by the variation in soil reaction or soil pH.

In general, bacteria prefer near neutral to slightly alkaline reaction between pH 6.5 and 8.0, the activity being sharply curtailed when the pH drops below 5.5. Fungi grow in acidic reaction between pH 4.5 and 6.5 and actinomycetes prefer slightly alkaline conditions.

Nitrification, nitrogen fixation, denitrification and other beneficial processes, carried out by different micro-organisms, are markedly influenced by the soil reaction or fluctuations of soil pH. Therefore, the soil pH regulates the qualitative composition of soil flora. All in all, a soil in the intermediate pH range presents the most satisfactory biological regime.

More Availability of Nutrients in relation to soil pH in Organic Soils

Nutrients	pH
Nitrogen	5.0 – 8.5
Phosphorus	4.5 – 6.5 8.5 – 9.0
Potassium	5.0 – 6.5 – 7.5
S	5.0 – 9.0
Ca	5.0 – 8.5
Mg	4.5 – 7.0
Mn	4.0 – 5.5 8.5 – 9.0
Fe	4.5 – 7.5
B	4.5 – 6.0 8.5 – 9.0
Cu	4.5 – 6.5
Zn	4.5 – 6.5
Mo	5.5 – 9.0

Relationship between pH of the mineral soil and availability of nutrient elements

Availability

Nutrients	pH
Nitrate	5.5 – 7.5
Potassium	5.5 – 7.0
Phosphate	5.5 – 7.0
Ca	6.5 – 8.0
Fe	4.5 – 5.5
Mg	5.0 – 7.5
S	5.0 – 8.0
Mn	4.5 – 5.5
Mo	6.0 – 8.0
Zn	4.5 – 6.0
Cu	4.5 – 7.0
B	5.0 – 7.5

BUFFERING OF SOIL

Buffer action

Buffering refers to resistance to a change in pH. If 1 ml of 0.01 N HCl is added to 1 litre of pure distilled water of neutral pH; the resulting solution would have a pH of about 5.0. Now on the other hand if small amount of acid is added to a soil suspension the resulting change in pH would be very small. The soil suspension is buffered against a change in pH. The power to resist a change in pH is called "buffering action" Buffer solution contains reserve acidity and alkalinity and does not change pH with small addition of acids or alkalis.

Buffering Capacity of Soils

Soils also exhibit buffering action. In order to produce a certain change of pH, it is usually necessary to add much more acid or alkali than the amount of H^+ and OH^- ions present in the soil suspension. This buffering action is due to the influence of weak acids and their salts soil behaves like a buffered weak acid and that will resist sharp changes in pH. This buffering mechanism may be explained as follows:

$$Al^{3+} + H_2O \rightleftarrows AlOH^{2+} + H^+$$

If these H^+ ions are then neutralized by the addition of small amounts of a base and the aluminium ions in solution are precipitated as Al $(OH)_3$ the equilibrium of the system, hence the pH, will tend to be maintained by the movement of adsorbed aluminium ions to the soil solution. These aluminium ions in turn hydrolyse, producing H^+ ions, and the pH tends to remain as it was before the addition of the base. As more base is added, then active H^+ ions are neutralised. In order to bring equilibrium, some of the H^+ ions held by the clay (potential acidity) will dissociate and cause slight increase in the concentration of H^+ ions in the active acidity and thereby resulting slight changes in pH. As the process is continued, less and less of the hydrogen tends to ionize during neutralization process and thereby resulting a gradual decrease in the active acidity (acidity develops due to concentration of H^+ ions in soil solution) with a constant increase in soil pH.

On the other hand, if H^+ ions are added to the system some of the additional H^+ ions will be adsorbed by the clay with the release of other ions. As a result, there will be a slight increase in the activity and corresponding decrease in soil pH. So because of the tendency of the soil colloid either to adsorb H^+ and Al^{3+} ions from soil solution or to give up those to the soil solution to maintain the equilibrium, the soil tends to resist sudden changes in pH and this action is said to be the buffering action of the soil.

Factors Affecting Buffering Capacity of Soil

There are various factors which affect the buffering capacity of soils, the most important of amount and kind of clay, organic matter content, CEC, CO_3, HCO_3 phosphate content and organic acid etc.

Soil containing large amounts of clay and organic matter are said to be highly buffered and require larger amounts of lime for affecting a certain change in pH than an acid soil containing smaller amounts of clay and organic matter. Keeping all other factors constant, the buffering capacity of a soil increased with increasing CEC of a soil.

Soils containing large amounts of calcium carbonate, organic acids and phosphates shows higher buffering capacity.

Soils containing large amount of 1:1 type clays are generally less strongly buffered than soils in which the predominant clay minerals are of the 2:1 type.

Buffering of Soils in Agriculture

The importance of buffering of soils is mainly two fold:

i) The stabilization of soil pH and

ii) The amounts of amendments necessary to affect a certain change in soil reaction or soil pH.

An abrupt change in pH causes a radical modification in soil environment and if it fluctuates too widely, the plants and micro-organisms would suffer seriously. Changes in soil reaction not only have a direct influence on the plants but also exerts an indirect influence on soil environment by creating sudden changes in the availability of nutrients. Deficiencies of certain plant nutrients and excess of others in toxic amounts would seriously upset the nutritional balance in the soil.

The amounts of amendments necessary to affect a given alteration in soil reaction also relate to the effectiveness of pH stabilization. The greater the buffering capacity of the soil, the larger must be the amounts of lime or sulphur to affect a given change in pH.

Soil Acidity

Chapter FOUR

Soil acidity is caused by ionizable hydrogen ions (Acid soil is a base unsaturated soil which has got enough of adsorbed exchangeable H^+ ions) so that to give soil a pH of lower than 7.0.

Soil acidity is common in all regions where precipitation (rainfall) is high enough to leach appreciable quantities of exchangeable base forming cations (Ca^{++}, Mg^{++}, K^+ and Na^+) from surface layers of soils.

DEVELOPMENT AND FORMATION OF ACID SOILS

- Continuous of use of acid forming fertilizer
- Acidc soils develop from acidic parent materials
- Most soils develop acidity by leaching (Al replacing the cation exchange sites). The more the leaching the soil become more acidic.

Nature of Soil Acidity

Two types according to the presence of H^+ ions

a) Active acidity : H^+ & Al^{3+} ions remain freely present in soil solution and produce limited acidity.

b) Reserve or potential acidity

In soil colloidal complex, exchangeable $H^+ Al^{3+}$ ions are adsorbed and give soil a pH lower than 7.0. The magnitude of this type of acidity is very high.

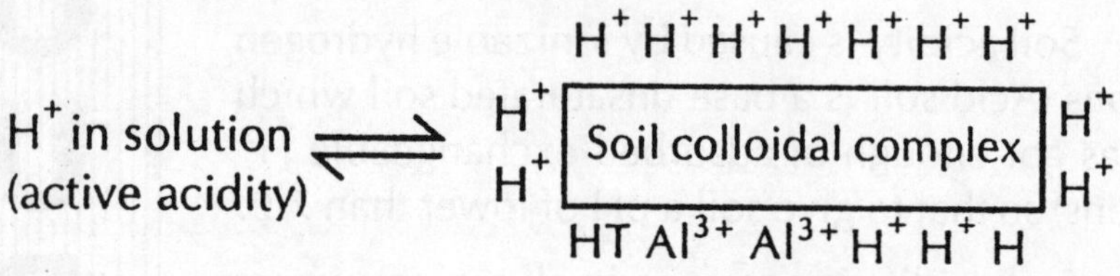

(Reserve Acidity)

Organic matter produces organic acids like citric acid, oxalic acid, acetic acid etc and imparts acidity to the soil. In mineral soils, acidity may be produced by the presence of free H_2SO_4

Effects of soil acidity on plants

1) Direct effect : (a) Toxic effect of H^+ ions on root tissues (b) influence of soil acidity on the permeability of the plant membrane for cations (c) disturbance in the balance between basic and acidic constituents through the roots.

2) Indirect effect : (a) Adverse effects on the availability of various nutrients i.e. P, Cu, Zn, etc. (b) Al, Mn, Fe become available in toxic amount (c) activities of soil micro-organisms adversely affected

The sources of H^+ ions — which are the initial source of solution acidity are as follows :

1. CO_2 from humus decomposition and root respiration

2. Oxidation of NH^{+}_{4} from fertilizers
3. Oxidation of added elemental sulphur
4. Excreted H^{+} ions by plant roots
5. Acid rain (sulphur and nitrogen oxide pollutants atmosphere)
6. Crop removal of the basic cations (Ca, Mg K, Na) and excretion of H by roots.

1. Carbondioxide from decomposing organic matters and root respiration dissolves in water to form weak carbonic acid

$$CO_2 + HOH \rightarrow H_2CO_3 \rightarrow HCO_3^- + H^+ \downarrow$$

2) Oxidation of NH_4 from fertilizers : Oxidised by bacteria to form nitrate and H^+ ions

$$NH_4^+ + 2O_2 \xrightarrow[\text{bacteria}]{\text{nitrifying}} NO_3^- + H_2O + 2H^+ \downarrow$$

3) Oxidation of added elemental sulphur

$$2S + 3O_2 + HOH \xrightarrow[\text{Bacteria}]{\text{Thiobacillus}} 2SO_4^{2-} + 4H^+ \downarrow$$

4) Some H^+ ions are also released by plant roots

- Measured by water cultures

In soils studies have shown pH values as much as 1.2 pH units lower in soil near roots than in the general mass of soil.

5) Acid rain — Major pollutant in Canada, Norway and Sweden. Original sources of these polluting oxides (sulphur oxide and nitrogen oxide) are burning of fossils fuels, such as wood, coal, petroleum products and forest and range fires. Acid rainfalls when air-borne sulphur oxide (mostly sulphur

dioxide SO_2) and nitrogen oxide (mostly nitric oxide, NO) are converted to sulphuric acid (H_2SO_4) and nitric acid (HNO_3) through oxidation and dissolved in rain drops. The amounts of sulphur dissolved in rainfall. Yearly range from 10 kg/ha, for light industrial areas to 30 kg/ha and as much as 100 kg/ha for soils in more industrialized states and near large plants, respectively.

6) Crop removal helps to make soils more acidic by depleting the reserves of Ca, Mg, K.

Two adsorbed cations – Hydrogen and Aluminium are largely responsible for soil acidity

$$\boxed{\text{Micelle}}\ Al^{+++} \longleftrightarrow Al^{3+}$$

Adsorbed alluminium (ion in soil solution)

Al ions in the soil solution hydrolysed

$Al^{3+} + H_2O \rightarrow Al(OH)^{2+} + H^+$ - released lower the pH of the soil solution

Classification of Soil Acidity

Three kinds of acidity:

a) Active acidity – due to the H^+ ion in the soil solution.

b) Salt replaceable acidity represented by the H^+ and Al^{3+} that are easily exchangeable by other cations in a simple unbuffered salt solution such as KCl.

c) Residual acidity — which can be neutralized by lime stone or other alkaline materials. All added up to the Total acidity of soil = (a) + (b) + (c)

Buffering Capacity of Soils

Buffering soils : There is a distinct resistance to change

in pH of the soil solution. This resistance is called buffering

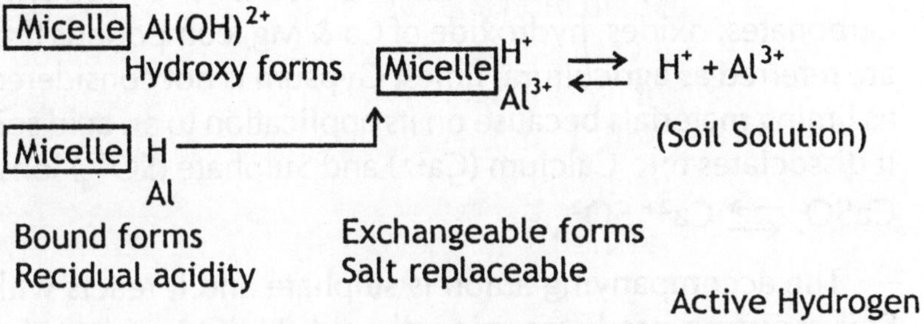

If just enough lime (base) is applied to neutralize the H^+ ion in solution. The result is resistance to change in the soil solution pH. In other words buffering effect is being exercised. H^+ and Al^{3+} ions together with the adsorbed base forming cations, largely control buffering just as they indirectly control the pH of the soil solution.

Buffer capacity of soils : The higher the exchange capacity of a soil, the greater its buffer capacity, other factors being equal. The higher the clay and organic matter contents, the more lime is required for a given change in soil.

Importance of Buffering to Agriculture: Changes in soil reaction (pH) have a direct influence on the plants and it also affects the availability of plant nutrients. Deficiency of certain plant nutrients and excess of others in toxic amounts would seriously upset the nutritional balance in the soil. Buffering prevents sudden changes and fluctuation in soil pH. So it regulates the availability of nutrients and also checks direct toxic effect to plants.

Reclamation of Acid Soils : Decreasing soil acidity use liming material (Management of acid soil)

1. Addition of lime

Soil acidity is commonly decreased by adding carbonates, oxides, hydroxide of Ca & Mg, compounds that are referred as agricultural limes. Gypsum is not considered as liming materials because on its application to an acid soil it dissociates into Calcium (Ca^{2+}) and Sulphate (SO^{2-}_4) ions.

$CaSO_4 \rightleftarrows Ca^{2+} SO^{2-}_4.$

The accompanying anion is sulphate and it reacts with Soil moisture produces mineral acid (H_2SO_4) which also increases soil acidity instead of reducing soil acidity.

$SO^{2-}_4 + H_2O \rightarrow H_2SO_4$ (Strong mineral acid).

Besides this, Calcium (Ca^{2+}) in the gypsum after dissociation will result in replacement of adsorbed aluminum (Al^{3+}) in a localized soil zone (when gypsum applied as band placement) with a significant lowering of soil pH. Therefore, gypsum doest not qualify as a liming material.

2. Liming materials

A) Carbonate forms lime are

(1) Marl, oyster shells, basic slag, and precipitated carbonates.

(2) Ground lime stone – liming material widely used. Minerals include lime stone – Calcite – $CaCO_3$ dolomite $Ca\,Mg\,(CO_3)_2$ effective in increasing crop yields.

B) Oxide forms : Commercial oxide of lime – burned lime, quick lime (Heating of lime stone in large commercial kilns).

Reaction

$$CaCO_3 + Heat \rightarrow CaO + CO_2 \uparrow$$

$$CaMg(CO_3)_2 + Heat \rightarrow CaO + MgO + 2CO_2 \uparrow$$

oxide of lime is more costly than lime stone.

3. Hydroxide form – lime – referred as hydrated lime

Adding water to burned lime

$$CaO + MgO + 2H_2O \; 0 \; Ca(OH)_2 + Mg(OH)_2$$

Hydroxide of lime appears on the market as a white powder. Coral shell, chalk, wood ash, press mud, by product material of paper mils, sugar factories, fly ash & sludge etc. are considered as liming materials also used for the amelioration of soil acidity.

Lime requirement of Acid Soils

The amount of liming materials required to bring about a desired pH. Change in determined by several factors including (a) the change in pH required (b) the buffer capacity of the soil (c) the chemical composition of the liming materials used (d) finess of the liming materials

Overliming : due to excess application of liming material that result pH values will be too high for optimal crop growth.

Excess lime include (1) deficiencies of iron, Mn, Cu, Zn (2) Reduced the availability of phosphate (3) restraints on the uptake of boron

Overliming should be avoided.

Acid Sulphate Soils

Soil with sufficient sulphides (FeS_2 and others) to become strongly acidic when drained and aerated enough for cultivation are termed acid sulphate soils or as the Dutch refer to those soils Cat clays. When allowed to develop acidity, these soils are usually more acidic than pH 4.0. Before drainage, such soils may have normal soil pH and are only potential acid sulphate soils. Generally acid sulphate soils are found in coastal areas where the land is immersed by salt water. In India, acid sulphate soil is mostly found in Kerala, Orissa, Andhara Pradesh, Tamil Nadu and West Bengal. The area covered under acid sulphate soils in Thailand and India combinedly is about 2 m. acres.

Formation of Acid Sulphate Soils

Lands immersed with waters that contain sulphates, particularly salt waters, accumulate sulphur compounds, which in poorly aerated soils are bacterially reduced to sulphides. Such soils are not usually very acidic when first drained in water. When the soil is drained and then aerated, the sulphide (S^{2+}) is oxidized to sulphate (SO^{2-}_4) by a combination of chemical and bacterial actions, forming sulphuric acid (H_2SO_4). The magnitude of acid development depends on the amount of sulphide present in the soil and the conditions and time of oxidation. If Iron pyrite (FeS_2) is present, the oxidized iron accumulates the acidity in but not as much as aluminium normal acid soils because the iron oxides are less soluble than aluminium oxides and so hydrolyze less.

Reactions Involving the Formation of Acid Sulphate Soils

Acid sulphate soils form due to oxidation of sulphides in soils. The slow oxidation of mineral sulphides in soils is non-biological unit soil pH reaches an acidity of pH 4.0, but the process is not well understood. Below pH 4.0, the bacteria, *Thiobacillus ferroxidans* are the most active oxidizer and activity builds up rapidly.

$$2S + 3O_2 + 2H_2O \xrightarrow{\text{Microbial oxidation}} H_2SO_4$$

(Elemental sulphur) (Sulphuric acid)

$$H_2S + 2O_2 \xrightarrow{\text{Microbial oxidation}} H_2SO_4$$

(Hydrogen sulphide) (Sulphuric acid)

Non-biological

$$2FeS_2 + 2H_2O + 7O_2 \rightarrow 2FeSO_4 + 2H_2SO_4$$

(Iron pyrite) (Ferrous sulphate) (Sulphuric acid)

Accelerated by bacterial (Thiobacillus ferroxidans)

$$4FeSO_4 + O_2 + 2H_2SO_4 \rightarrow 2Fe_2(SO_4)_3 + 2H_2O$$

(Ferrous sulphate) (Ferric sulphate)

Rapid in acid pH (non-biological)

$$FeS_2 + 7Fe_2(SO_4)_3 + 8H_2O \rightarrow 15FeSO_4 + 8H_2SO_4$$

(Ferrous sulphate)

Classification and Characteristics

Acid sulphate soils contain a sulphuric horizon which has a pH of the 1:1 soil: water ratio of less than 3.5, plus some other evidences of sulphide content (yellow colour).

Sulphaquepts, Sulphihemists, Sulphohemists and Sulphaquents great groups are included in acid sulphate soils.

Such strong acidity in acid sulphate soils results toxicities of aluminium and iron, soluble salts (unless leached) manganese and hydrogen sulphide (H_2S) gas. Hydrogen sulphide (H_2S) gas. Hydrogen sulphide (H_2S) often formed in low land rice soils causing akiochi disease that prevents rice plant roots from absorbing nutrients.

Management of Acid Sulphate Soils

Management techniques are extremely variable and

depend on many specific factors viz., the extent of acid formation, the thickness of the sulphide layer, possibilities of leaching or draining the land etc. The general approaches for reclamation are suggested below :-

i) Keeping the area flooded

Maintaining the reduced condition of flooded (anaerobic) soil inhibits acid development, which requires oxidation. This solution almost limits the use of the area to rice growing. Unfortunately, droughts occur and can in short time periods cause acidification of these soils. The water used to flood the potential acid sulphate soils often develop acidity and injure crops.

ii) Controlling water table :

If a non-acidifying layer covers the sulphuric horizon, drainage to keep only the sulphuric layer under water (anaerobic) is possible.

iii) Liming and leaching

Liming is the primary way to reclaim any type of acid soil. Acid soil may require 11-45 M.T/ha of lime in a 20 year period whereas acid sulphate soils may require from several metric tons/ha/yr upto even 224 MT/ha (100 t/Acre) within a 10-year period or less. If these soils are leached during early years of acidification lime requirement are lowered.

Chapter FIVE

Chemistry of Submerged Soils

DEFINITION

Submerged soils are soils that are saturated with water for sufficiently long time in a year to give the soil the following distinctive gley horizons resulting from oxidation – reduction processes :

i) a partially oxidized "A" horizon high in organic matter

ii) a mottled zone in which oxidation and reduction alternate and

iii) a permanently reduced zone which is bluish green in colour

The soil is intermittently saturated with water, oxidation of organic matter is slow and it accumulates in the 'A' horizon. In the 2nd horizon Fe and Mn are deposited as a rusty mottles or streaks if the diffusion of O_2 into the soil is slow, if the diffusion is rapid, they are deposited as concretions.

Kinds of submerged soil

There are different kinds of submerged soil, of which (i) continuous submerged soils and (ii) alternate submerged soils are important for rice cultivation.

Continuous submergence at a static 2.5 – 7.5 cm depth provides the potential to produce optimum rice yields. Generally continuous submergence at 15 cm depth or more has the potential to produce yield similar to those at 2.5 cm water depth. However, in some dry seasons a 15 cm depth or more may reduce grain yield.

Importance : Rice is the only major food crop that can be grown under various degrees of submergence. Submerged conditions exhibit the following characteristics (i) greater amount of soil solution (ii) reduced oxygen level (iii) a reduced aerobic microbial activity and (iv) an altered chemical status of the soil.

The rice plant absorbs nutrients and grows under these specific complex conditions.

Generally systems of rice cultivation in India and especially in West Bengal can be classified into two major types (i) Semi-dry system – where rice is grown primarily as a rainfed crops especially in uplands. The yields under such conditions are normally low (ii) wet system, practised in low lands or where irrigation is available. In this system of rice cultivation, the production is very high. Soil submergence for rice cultivation is thus essential. Because as a result of soil submergence a variety of changes like physical, chemical, electrochemical and biological properties are modified of which majority of soil properties is suitable for rice cultivation.

Properties of submerged soil

There are various changes like Physical (depletion of oxygen, accumulation of CO_2 compaction, bulk density, puddling, gaseous exchange and movement of water etc.)

Chemical (Soil reduction and transformation of different nutrient elements etc)

Electrochemical (Soil pH, specific conductance and redox potential, Eh etc) and Biological properties (decomposition of organic matter, mineralization and immobilization processes etc) of soils that are strongly influenced by the soil submergence.

PHYSICAL PROPERTIES

1. Diffusion of molecular oxygen and development of aerobic/anaerobic layer

When a soil is submerged, water replaces the air in the pore spaces. Except in a thin layer at the soil surface, and sometimes a layer below the plough sole, most soil layers are virtually oxygen free within a few hours after submergence. The oxygen diffusion in the water layer above the soil is very slow and the rate of O_2 consumption in reduced soil is high. Because of this high demand O_2 in submerged soil and slow O_2 takes place within a day or so of submergence. Some O_2 trapped in blocked pore spaces is rapidly utilized by facultative anaerobic organisms.

The greater potential consumption of O_2 as compared to the available supply through the flood water results in two distinctly different layers being formed in submerged soil an oxidized or aerobic surface layer where O_2 is present and a reduced or anaerobic layer in which no free O_2 is present Fig. 16 (i) & (ii).

Oxygen reaching the submerged soil surface is utilized in several ways:

a) as an electron acceptor by soil micro-organisms in their cell respiration,

b) in the chemical oxidation of ferrous iron (Fe^{2+}) and manganous manganese (Mn^{2+}), and

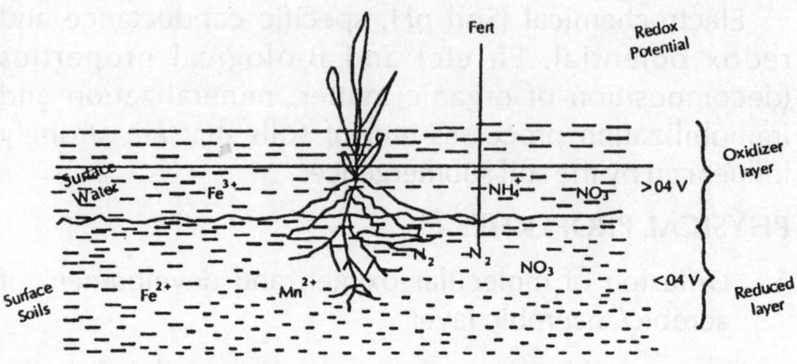

Fig. 16(i): Soil profile of submerged soil

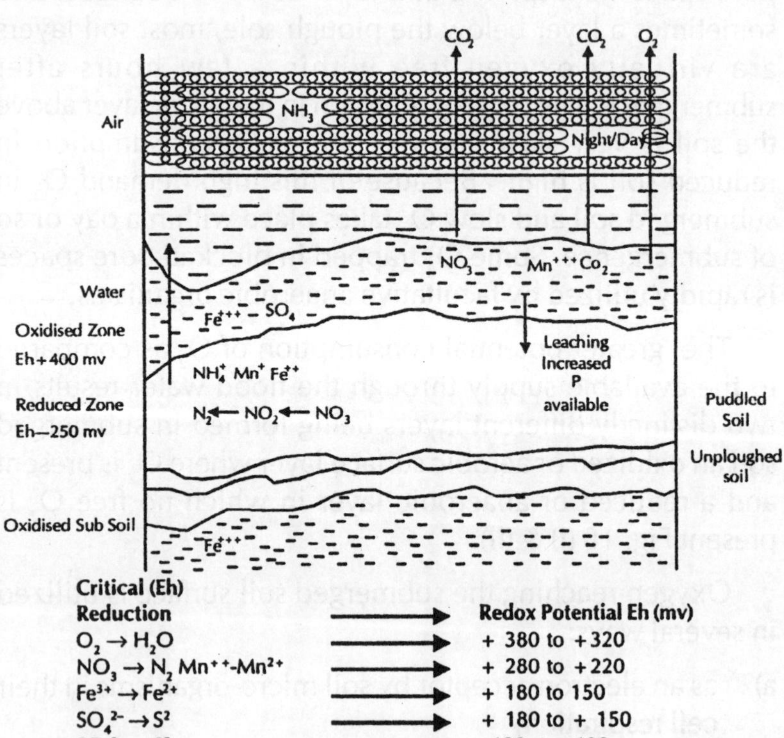

Critical (Eh) Reduction		Redox Potential Eh (mv)
$O_2 \rightarrow H_2O$	→	+ 380 to + 320
$NO_3 \rightarrow N_2$ Mn^{++}-Mn^{2+}	→	+ 280 to + 220
$Fe^{3+} \rightarrow Fe^{2+}$	→	+ 180 to 150
$SO_4^{2-} \rightarrow S^2$	→	+ 180 to + 150
$SO_4^{2-} \rightarrow S^2$	→	–120 to –180
$CO_2 \rightarrow CH_4$	→	–200 to –280

Fig. 16(ii): Oxidised reduced zones in submerged soil

c) for biological oxidation, such as oxidation of NH_4^+ - N to NO_3^{3-} - N and S^{2-} and S^o to SO^{2-}_4.

2. Aeration status of a submerged soil

Immediately after submergence, the normal process of gaseous exchange between soil and air is restricted. The net result is that the concentration of O_2 in the soil is reduced to a very low value, while that of soil gases, notably CO_2 is increased, especially if conditions are favourable for biological activity.

3. Accumulation of CO_2

Soil gases like CO_2 and methane (CH_4) accumulate due to submergence and also may escape as bubbles if pressure builds up. It has been recorded that during the first 3 weeks submergence some soils may generate CO_2 upto 2.5 t/ha.

Due to presence of large amounts of carbonic acid (H_2CO_3) the chemical equilibriaum in soil water system containing high amount of Ca^{++}, Mg^{++}, Fe^{++} and Mn^{++} affects.

Soil compaction and puddling

Soil Compaction

In compaction, soil solids are rearranged with compression of liquid and gaseous phases accompanied by volume change. Soil compaction affects the water retention characteristics, water intake rates, and gaseous exchange. In compacted soil, bulk density, microvoids, thermal conductivity and diffusivity and nutrient mobility increases and on the other hand, macrovoids, hydraulic conductivity and water intake rates decreases. Generally medium textured soils are most susceptible to compaction Therefore, it may be concluded that soil compaction cuts down percolation losses and reduces the water requirement of rice, appears to be more practical & economical tillage, practical than puddling for increasing rice yield & water use efficiency.

4. Puddling

Puddling refers to breaking down soil aggregates at near saturation into ultimate soil particles. The mechanical reduction in the apparent specific volume of soil is found due to puddling. Only soils with more than 20% of clay particles are prone to puddling.

Puddling and subsequent flooding differentiate low and rice soils chemically and pedologically from other arable soils. An important difference is between a dry land and puddle low land soil system. Puddling, an intensive wet land cultivation, breaks the natural aggregates to finer fractions. It decreases the apparent specific volume and hydraulic conductivity, creates an anaerobic environment and affects redox (Eh) and pH values.

Puddling influences physical, chemical and biological soil properties which in turn influence rice growth. Puddling has both short and long-term effects on soil and rice growth.

Short term effects :

(i) **Soil Structure:** Puddling destroys aggregates and peds. Wetting dry soils cause uneven swelling in aggregates and explosion of trapped. The aggregates are slaked. Aggregates, depending on their stability.

(ii) **Bulk density and Soil Strength:** The effect of pudding on BD depends on soil aggregation before puddling.

(iii) **Soil porosity:** The effect of puddlign on perosity depends on the soil particle orientation in the puddled layer. If a more open structure results from puddling, total porosity will increase.

(iv) **Gas exchange:** Gas of O_2 between soil and atmosphere is strictly restricted. O_2 concentration decreased & CO_2 concentration increases.

Long term effects. Long term puddling forms a hardpan in the subsoil below the puddled layer. It is due to physical compation and precipitation of Fe, Mn & Si. Chemically cemented pans are formed in the oxidised subsoil (15-20 cm deep). Ferrolysis is another long term effect of puddling that may lower soil productivity. Ferrolysis is the process of clay decomposition and transformation under the influence of periodic reduction of iron oxides to ferrous ions (Fe^{2+}).

Ferrolysis during Wet Seasons :

Water Saturation : (1) Reduction (2) Fe2+ production from freeiron oxides.

(1) Reduction H^+ consumption Formation of Soluble Al-polymoers pH rise decrease in Fe^{2+} activity formation of incomplete - Al interlayers in 2:1 clay mineral trapping of formerly exchangeable - Fe^{2+} in interlayers.

(2) Fe^{2+} production Displacement of exchangeable bases and Al by Fe^{2+} Eluviation of basis, Al, Fe^{2+}

Ferrolysis during Dry season :

Air entry oxidation of exchangeable (& disoolved) Fe^{2+} to free iron oxides H^+ production Formation of H^+ clay pH decreases Clay decomposition of Al, Mg Clay liberation of Secondary silica formation of Secondary silica aggregates.

Electrochemical changes in submered Soils are (i) decrease in redox potential (Eh) (ii) increase in pH of acid soil & decrease in pH of alkaline soils, (iii) increase in specific conductance (iv) Cation exchange reactions involving Fe^{2+} (v) Sorption & desorption of ions.

- Low redox potential harms germination and seedling emergence.
- Low redox potential results net increase in the soil nitrogen regime.

  (Destroys No3-N, favours NH_4^+-N accmulation)
- Low redox potential helps in (rice) incrreaseing the availability of N, P, Si, Fe, Mn and Mo
- Redoxpotential causes harmful effect to rice by decreasing the availability of S, Cu and Zn.

Chapter SIX

Soil Salinity and Alkalinity

SALINE AND ALKALINE SOIL

Saline and alkaline soils are those that have an alkaline reaction or whose pH is greater than 7.0. This is due to the presence of excess of sodium salts or to the predominance of sodium among exchangeable bases.

Saline and alkali soils occur most commonly under arid climate.

Area under saline and alkaline soils (lakh hactares)

Sl. No.	State	Area	Sl. No.	State	Area
1.	Uttar Pradesh	12.95	9.	Karnataka	4.04
2.	Gujarat	12.14	10.	Madhya Pradesh	2.24
3.	West Bengal	8.50	11.	Andhra Pradesh	0.42
4.	Rajasthan	7.28	12.	Delhi	0.16
5.	Punjab	6.89	13.	Kerala	0.04
6.	Maharashtra	5.34	14.	Bihar	0.04
7.	Haryana	5.26	15.	Tamil Nadu	0.04
8.	Orissa	4.04		Total	69.49

Total 69.49 lakh ha or 7.00 million ha.

Nature and Classification

The presence of an excess of Na salts and the

predominance of sodium in the exchangeable complex are divided into two main groups. (1) Saline Soils (2) Alkaline soils (sodic soils)

1) Saline Soils : Saline soils contain an excess of Na salts, but its colloidal material is not yet sodiumised (pH less than 8.5, ESP less than 15, EC is more than 4 dS/m.)
2) Alkali soils : (Sodic soils) In the case of alkali soils, the exchangeable complex contains appreciable quantities of exchangeable Na. Such soils may or may not contain excess salts (ESP greater than 15)

Alkaline soils (pH greater than 7.0)

↓

Saline soils pH less than 8.4
ESP less than 15 EC is more than 4 dS/m.

Alkali soils (sodic)
(ESP greater than 15)

↓

Non-saline alkali soils
EC less than 4 dS/m
pH more than 8.5
↓
Degraded alkali soils
$CaCO_3$ Absent
Horizon A & B - Acidic
pH >8.5
Dark coloured
Compact soils

Saline alkali soils
pH - 8.5
EC greater than 15
EC greater than 4 dS/m

Characteristics of Saline and Alkaline Soil

1. Saline Soil : When the soil contains excess of Na salts and clay complex still contains exchangeable Ca, the soil is known as saline soil or white alkali or brown alkali soil. The process of accumulation of salts leading to the formation of soils is known as salinization.

i) Saline soils contain usually Cl, SO_4, HCO_3 and some time NO_3 of Na. The presence of Cl and SO4 of Na give a white colour on the soil surface when nitrates are in excess they give a brown colour to the soil.

ii) ESP is very low, being less than 15% of the total CEC.

iii) As a consequence of low ESP, generally pH varies between 7.5 and 8.5.

iv) Total soluble salt content is more than 0.1%. It is high enough to interfere with normal growth of most plant species.

v) EC of solution is > 4 dS/m.

vi) Saline soils usually have a surface crust of white salts, especially in the dry season when the net movement of soil moisture is upward. Salts dissolved in the soil water move the surface where they are left as a crust when the water evaporates.

2. Alkali soils (Sodic soils)

These soils contain sodium carbonates (Na_2CO_3) in abandance. These soils are often called "black alkali" soils because they are black, owing to the effect of the high Na content which causes the dispersion of the organic matter. These soils are also called typical usar soils.

i) ESP > 15%

ii) pH ranges 8.5 to 10 (strongly alkaline)

iii) Total soluble salt (Na) content is less than 0.1%

iv) EC is less than 4dS/m

v) Colloidal complex is deflocculated and dispersed. The clay swells and chokes the soil pores. Hence, the permeability to water and air is poor (or infiltration or aeration is poor).

vi) The presence of free Na carbonate has a toxic effect on plant roots. Also, the high pH and poor physical condition of soil adversely affect plant growth.

vii) Sodium carbonate absorbs organic matter so there is great depletion of organic matter. Therefore, these soils are almost barren (Usar).

Formation of saline and alkaline soil

Origin or development of saline and alkaline soil depends upon following factors (Saline and alkali soils occur most commonly under arid climate):

i) Arid and semi-arid climate : Due to low rainfall and high evaporation soils are in alkaline in reaction, the presence of sodium salts. Low rainfall regions is not sufficient to leach out the soluble products of weathering and hence the salts accumulate in the soil. During rain, the salts dissolve in rain water and move down in the lower layers. In dry weather the salts move up with the water and are brought up to the surface where they are deposited as the water evaporates.

ii) Poor drainage : During the period of high rainfall, the salts are leached from the upper layer and, if the drainage is impeded, they accumulate in the lower layer. When water evaporates, the salt is left in the soil. Such soils are generally developed in low lying areas or in basin shaped areas.

iii) High water table: The ground waters of arid regions usually contain considerable quantities of soluble salts. If the water table is high, large amounts of water move to the surface by capillary action and evaporated, leaving soluble salts on the surface.

iv) Overflow of sea water over lands : Low lying areas near the sea which get sea water during tides, salt water accumulates and enrich the soils with salts.

v) Introduction of Irrigation Water : The ground water of arid regions are generally saline in nature. With judicious irrigation the percolating water may get linked with the saline ground water. During the dry weather the soluble salts of the ground water may, thus get carried to the surface and increase the salinity of the land. The irrigation water may be itself rich in solubles salts and add to the salinity of the soils.

vi) Salts blown by wind : In arid regions near the sea, lot of salt is blown by wind year after year and get deposited on the lands. Due to low rainfall they are not washed back to sea and thus add salinity to the land. The salinity of Rajasthan has developed to a great extent, due to this reason.

vii) Saline nature of parent rock material : If soils develop from saline nature of parent rock materials. Soils would be saline.

viii) Excessive use of basic fertilizers : Use of alkaline fertilizers, like sodium nitrate, basic slag, etc. may develop alkalinity in soil.

ix) Humid and semi-humid regions : Alkaline soils develop in other area also, eg. in semi-humid and temperate regions, especially in depressions where drainage is defective and where the underground water table is high or close to the surface.

EFFECTS OF SOIL SALINITY AND ALKALINITY

Saline Soils

i) Absorption of water and nutrients : Excessive salts in the soil solution increase the osmotic pressure of soil solution in comparison to cell sap. This prevents absorption of moisture and nutrients in adequate amounts by the roots.

ii) Salt toxicity : When the concentration of soluble salts increase to high level then it produces toxic effect directly to plants. Saline soils are usually barren but potentially productive soils.

Alkali Soil (Sodic soil)

i) Dispersion of soil particles : Under alkali (Sodic) soil conditions, the drainage is not due to salt concentration. The Na adsorbed by clay and colloids causes dispersion of clay which results in a loss of desirable structure and development of compact soil.

ii) Physical properties affected : Due to compactness of soil aeration, permeability, drainage and microbial activity are reduced.

iii) Availability of plant nutrient reduced : The high pH in alkali soil causes a reduction in the availability of plant nutrients such as P, Ca, N, Fe, Cu, Mn, Zn.

Reclaimation of Saline and Alkali Soils

Schoonover (1959) in his study of soil problems in India, has listed the following technical requirements for reclaimation of saline and alkali soils:

1. Adequate drainage
2. Availability of sufficient water to meet crop use and also leach the salt below the root zone in the soil
3. Better than average soil management to include perfect land levelling, good bunding for irrigation and advanced agronomic practices.
4. Protection and reclamation to be taken in large blocks
5. Irrigation water should be of good quality

Saline Soil Reclamation and Management

Saline soils in which the soluble salts contain appreciable amounts of Ca^{++} and Mg^{++} do not develop into alkali soils by the action of leaching water. The reclamation is comparatively easy in such soils. The main problem is to leach the salts downward below the root zone and out of contact with subsequent irrigation water. Following methods may be used for removal of salts:

A) Mechanical Methods

(i) Flooding and leaching down of the soluble salts : The leaching can be done by first ponding the water on the land and allow it to stand therefore a week. Most of the soluble salts would leach down below the root zone. After a week standing water (dissolved with soluble salts) is allowed to escape. Such, 2 to 3 treatments are given to reclaim highly saline soils. Sometimes gypsum is also added to flood water when the soluble salts are low in calcium to check development of alkalinity.

(ii) Scraping of the surface soil : When the soluble salts accumulate on the soil surface, scraping helps to remove salts. This is a temporary cure and salinity again develop on such lands.

B) Cultural Methods (Crop, Soil and Water Management)

i) Providing proper drainage : If the soil is not free draining artificial drains are opened or tile drains laid underground to help wash out the salts.

ii) Use of salt free irrigation water : Salt free good quality of irrigation water should be used.

iii) Proper use of irrigation water: It is known that as the amount of water in the soil decreases the concentration of salts in the soil solution increases, thus moisture should be kept at optimum field capacity.

iv) Planting or sowing of seeds in the furrow: The salt concentration even in smaller amounts, is most harmful to the germinating seedlings. Water generally evaporates from the highest surface by capillary and hence these points have maximum salt concentrations. If the seeds or seedlings are planted inside the furrows, they escape the zone of maximum salt concentrations and thus can germinate and develop properly during their early growth stage.

v) Use of Acidic Fertilizer : In saline soil, acidic nature of fertilizers (e.g. Ammonium sulphate) should be used.

vi) Use of organic manures : The organic manures have very high water holding capacity. When sufficient amount of these manures are added the WHC of soil increases and as a result the conductivity of the soil solution decreases.

vii) Ploughing and levelling of the land : Ploughing and levelling of the land increases the infiltration and percolation rate. Therefore, salts leach down to the lower levels.

viii) Retardation of water evaporation from soil surface: Water may be considered in the soil by retarding the water evaporation. Thus salts may remain in the lower levels with the water.

ix) Growing of salt tolerant crops :

(a) High salt tolerant crops : Barley, sugarbeet, cotton etc.

(b) Moderately salt tolerant crops : Wheat, rice, sorghum, main flax etc, safflower

(c) Low salt tolerant crops : Beans, radish, white clover etc.

Reclamation and Management of Alkali soils

(Saline alkali and non-saline alkali)

Management of alkali soils : Chemical method

I. Chemical Method

1. Application of Gypsum : $Na_2CO_3 + CaSO_4 \leftrightarrow CaCO_3 + Na_2SO_4$
2. Application of sulphur : $2S + 2H_2O + 3O_2 \rightarrow 2H_2SO_4$

$$Na_2CO_3 + H_2SO_4 \leftrightarrow CO_2 + H_2O + Na_2SO_4 \downarrow$$

3. Addition of organic matter
4. Addition of molasses

II. Cultural Method : Same as saline soils

Chapter SEVEN

Quality of Irrigation Water

The concentration and composition of dissolved constituents in a water determine its quality for irrigation use. Quality of water is an important consideration in any appraisal of salinity or alkali conditions in an irrigated area. All irrigation waters contain some salts, but the concentration and nature of salt vary. The quality of irrigation water depends primarily on the total amount of salt present and the proportion of Na^+ to other cations and certain other parameters.

Agricultural crops receive moisture for their growth and development mainly from two sources (i) rain water and (ii) irrigation water. Rain water is practically free from mineral matters, although it contains dissolved gases (N, Ar, O_2, CO_2 and NH_3) obtained from the atmosphere.

Irrigation water consists of (a) surface water (b) ground water

CRITERIA FOR EVALUATING OF IRRIGATION WATER

Water quality is determined according to the purpose for which it will be used. For irrigation waters, the usual criteria include salinity, sodicity (sodium content) and element toxicities. All these criteria are being discussed here. Besides, many important criteria in assessing water quality for other uses namely, taste, colour, odour, turbidity, temperature, hardness, pH, BOD (Biological oxygen demand) or COD (Chemical oxygen demand) nutrient contents like N, P etc. and other pathogenic organisms are sometimes but not usually important for irrigation water.

Sludge and manure effluents have high BOD and COD and are undesirable for aquatic life (fish, algae, etc). Such waters would also be more rapidly depleted of oxygen when added to poorly drained soils. High BOD waters when used on soil can cause poor aeration (inadequate oxygen) conditions faster than would low BOD water. When sewage and industrial effluents are used for irrigation purposes, pathogenic organisms causing various diseases to agricultural crops needs to be studied. Free of viable pathogenic organisms in the effluents should be used as an irrigation water.

IRRIGATION WATER RESOURCES

Water resource class

The water resources can be divided mainly in two classes

(i) Surface water

(a) Flowing Water (i) Rivers, canals, drains

(b) Stagnant water (i) Lakes, tanks, sea water

(c) Sewage water

(ii) Ground water

1. Rain Water

Factors affecting water resources

i) Climatic factors

a) Rainfall, intensity, duration and distribution

b) Snow

c) Evapotranspiration

ii) Physiographic factors

a) Basic characteristics

Geometric factors : drainage area, shape, slope and stream density

Physical factors :. land use, surface infiltration conditions. soil types etc.

b) Channel characteristics carrying capacity and storage capacity

iii) Geological factors

c) Lithologic including composition, texture, sequence of rock types and the thickness of rock formation.

Properties of waters from different sources

1. Rain water – Fresh quality water but its fertility is poor
2. River and canal waters are usually fresh and some time alkaline. Its fertility is good
3. Ground water may be either fresh, alkaline or saline
4. Sea water – highest in salinity

Nature and amount of soluble salts in natural waters

The nature and amount of soluble salts present in natural waters are affected by several factors:

1) Physico-chemical characteristics of the rocks with which water is coming into contact
2) Geological history of the area
3) Climate of the area including duration and frequency

of rain, humidity, intensity and duration of sunshine, wind velocity, etc.

4) Micro-organisms

5) Chemical, physical and mineralogical characteristics of the soil with which the water comes into contact either during surface runoff, or when it percolates into the underground reservoirs

6) Topography of the area

7) Plant or vegetation cover

8) Ground water mixing with waters of deep seated origin

9) Human interference in bringing about micro-changes in the hydrological cycle, use of water for irrigation purpose, use of fertilizers etc.

Irrigation Water Quality Criteria are required by

1. Geologists and water resources engineers who are responsible for the development of water resources for irrigation.
2. Irrigation authorities who are responsible to the farmer for the quality of water delivered to the crops.
3. Soil scientists who are responsible for the protection of the soil from becoming saline and/or alkali (sodic).
4. Agronomists who have to decide the agronomic and cultural practices for obtaining optimum crop production.
5. Irrigation engineers who have to decide the suitable irrigation practices.
6. Plant scientists who have to decide the crop and its

variety to be grown.

7. Drainage engineers who are responsible for the utilization of drainage waters.
8. Water pollution control authorities who must regulate the discharge of sewage or industrial waste to the rivers, lakes, and ground water basins.
9. Environmental planning and protection authorities who are responsible for the conservation of the environment.

Characteristics that determine Water Quality

The characteristics of irrigation water that appears to be the most important in determining its quality, however, depending on climatic conditions, irrigation practices, soil water retentivity characteristics crop tolerance, depth to water table and agronomic practices etc. are the following:

1. Salinity hazard (Total concentration of soluble salts) : Electrical conductivity (EC)
2. Specific ion-toxicity hazard (Ionic composition)
   a) Major constituents (Na, Mg, Cl, HCO_3 , CO_3 silica, nitrate)
   b) Minor constituents (Boron, Lithium, Flourine, and other micro toxicants.
3. Sodicity hazard (Relative proportion of sodium to other cations) : Sodium Adsorption Ratio (SAR)
4. Alkalinity hazard (Bicarbonate concentration as related to the concentration of Ca + Mg) : Residual Sodium Carbonate (RSC)

Combined evaluation of two parameters (Practical importance)

5. EC & SAR
6. SAR & RSC

1. Salinity hazard

The total concentration of soluble salts, is the single most important criterion which has been used conventionally for determining quality of irrigation water. It is measured quantitatively in terms of 'Electrical Conductivity' (EC) because this is very closely related with the sum of major cations (or anions) determined by chemical analysis

EC expressed in (μ mhos/cm) or (micromhos/cm) at 25°C

(mmhos/cm) or millimhos/cm at 25°C

(dSm^{-1}) or deci Simens per meter

In evaluating the quality of irrigation waters only on the basis of EC, it is presumed that low salinity waters have low RSC and high salinity waters have low SAR. Therefore, it is necessary that low salinity waters are evaluated for RSC and high salinity waters for SAR.

2. Specific ion toxicity hazard

The quality of irrigation water used to be evaluated with respect to Na on the basis of soluble Na percentage (SSP) calculate as below :

$$SSP = \frac{\text{Soluble Sodium Concentration (meq/lit)}}{\text{Total cation concentration (meq/lit)}} \times 100$$

SSP useful in characterizing water since a high value indicates a soft water and low value a hard water.

Both soils and plants are affected by high Na irrigation waters. No soils are relatively impermeable to air water. They are hard when dry difficult to tilth.

Mg. Abundant cation usually found in high salinity waters. In low salinity waters, Ca dominates over Mg ions. With an increase in EC of waters, Mg/Ca ratio tends to increase.

Most important criteria in determining quality of water for irrigation is the Mg content of the irrigation water.

A harmful effect on soil appears When the ratio $\frac{Mg}{Ca+Mg}$ exceeds 0.5

Cl : increases in natural waters with an increase in EC and Na ions. Therefore Cl are most dominant in very high salinity waters. Cl ions have effect on the physical properties of the soil. It causes specific ion toxicity to plants. Cl salts are more harmful than sulphates

Potential salinity = $Cl^- + \frac{1}{2} SO^-_4$ (meq/lit)

Limits potential salinity for soils	5-20 meq./lit (good)
	3-15 meq/lit (medium)
	3-7 meq/lit (low) (permeability)

3) Sodicity hazard

a) Sodium adsorption ration (SAR)

Apart from the total salinity, the next important consideration is whether the use of irrigation water of given quality will cause sodicity in the soil.

The quality of irrigation water should be such that the irrigated soil does not exceed ESP value of 15 and pHs value of 8.5

Sodic soil – SAR greater than 15, so the irrigation water which does not increase SAR of the soil beyond 15 is considered to be of good quality.

$$SAR = \frac{Na}{\left(\frac{Ca+Mg}{2}\right)^{1/2}}$$

High SAR waters cause permeability problem in the soil wherein the rate of water infiltration into and through the soil is reduced to such an extent that the crop is not adequately supplied with water and consequently the yield is reduced. SAR of the water normally increases with an increase in EC of the waters.

4. Alkalinity hazard

Residual Sodium Carbonate (RSC)

Residual bicarbonates reacts with Na and forms sodium bicarbonate which upon evaporation is converted into sodium carbonate sodium, carbonate reacts with normal soil and replace exchangeable Ca with Na.

RSC to be calculated as

$(CO_3 + HCO_3) - (Ca + Mg)$

meq/lit meq/lit

RSC

More than 2.5 meq/lit were not suitable for irrigation

Between 1.25 and 2.5 meq/lit as marginal

less than 1.25 meq/lit were probably safe.

CO_3 & HCO_3 : Occurs in low salinity waters and its concentration usually decreases with an increase in EC. The proportion of HCO_3 ions is higher than Ca ions have been considered undesirable because after evaporation of irrigation waters, H CO_3 ions tend to precipitate Ca ions

RSC $(CO_3 + HCO_3) - (Ca + Mg)$
meq/lit

residual sodium bicarbonate (RSBC) = $HCO_3 - Ca$
meq/lit meq/lit

The presence of CO_3 ions is least desirable in waters because they tend to virtually eliminate Ca ions from the water and cause strong alkalinisation in the irrigated soils.

Silica : Abundant element in igneous rock

The importance of silica has usually not been recognized in evaluating quality of waters for irrigation but it seems to be necessary in view of the fact that it tends to cause alkalinity hazard.

Nitrates : highly soluble, natural waters contain only small quantities, usually less than 10 meq/lit.

Nitrates in water may result from leaching of soils and rocks from fertilizers, normal decomposition of plants and animals, sewage, industrial effluents and living organisms.

Saline waters – 2 – to 5 meq/lit Nitrate cause beneficial effect.

Nitrates cause beneficial effect on the plant growth and help partly in the substitution of nitrogenous fertilizers.

About 10 kg N/ha is added to the soil/irrigation (7.5 cm depth) if irrigation water contains nitrates equal to 1.0 meq/lit > 10 meq/lit nitrates tend to cause specific ion toxicity on plant growth. Nitrates are more harmful than Cl.

Minor constituents

Boron : The concentration of boron ranges usually from traces to 5 meq/lit in natural ground water. However, in very high salinity waters some time it can occur up to 10 meq/lit. Boron increases with an increase in EC and SAR of the waters.

Boron in sea water 5 meq/lit

Salt affected soils 10 meq/lit

Normal soils below 2 meq/lit

Boron is essential to plant growth but may prove exceedingly toxic at concentrations slightly above optimum. Plants usually do not show symptoms of boron deficiency if irrigated with water containing at least 1 meq/lit of Boron, but injury may develop on the more sensitive plants, when irrigated with water containing boron in excess 3 meq/lit. The injury appears as a burning and browning of the leaf top followed by yellowing of the margins. Retard growth and premature sheading of leaves are other symptoms.

When a water containing boron is used for irrigation, part of it adsorbed by the soil with the balance remaining in the soil solution, depending upon texture, clay mineral type, organic matter, Ca CO_3 salinity and sodicity of the soil. Boron does not affect the properties of soil.

Adsorption of Boron as affected by texture of the soil.

B added meq/lit	Boron in solution	Boron adsorbed
	Sand Texture	
5	3.8	1.2
10	7.8	2.2
50	43.0	7.0
Sandy loam		
5	3.0	2.0

10	6.4	3.6
50	42.0	8.0
Clay (Medium Black Soil)		
5	1.2	3.8
10	3.8	6.2
50	25.0	25.0

Lithium : Lithium is present in some minerals but is not abundant in nature. The more common minerals which contain from 2% to 10% lithium oxide are petalite, spodumene, lepidolite, amblygenite.

Lithium is generally found in soils developed from mica minerals and is more frequent in arid regions. Surface water generally do not contain lithium more than 0.3 meq/lit. It increases with the increase in EC of waters. The effect of lithium is not known. So lithium in waters and soils is not required at present.

Fluorine : Calcium fluoride (fluorite) is the main source of fluoride. The element is often characteristics of waters from deep strata and is frequently found in salt water from oil wells surface waters generally do not exceeds 0.3 meq/lit. Flourine does not affect soil properties.

Other micro toxicants : Apart from Boron, lithium, fluorine, there are some more trace elements, which if present in irrigation waters may also cause a toxic effect

Aluminium (Al), Arsenic (As), Beryllium (Be), Cadmium (Cd), Cobalt (Co), Chromium (Cr), Copper (Cu) Lead (Pb), Lithium (Li), Mn, Fe, Mo, Nickel (Ni), Selenium (Se), Vandaium (V) Zn.

New Classification of waters

With respect to salinity hazard on the basis of EC, the irrigation waters may be classified into six salinity classes

i) Salinity (EC)

a) C-O Non saline water (EC < 0.2 dS/m) may create a severe permeability problem in the soil because infiltration rate into the soil is adversely affected due to lack of salts in the water to such an extent that the crop is not adequately supplied with water and yield is reduced

b) C-1, Normal waters (EC 0.2 – 1.5 dS/m)

c) C-2 Low salinity waters (EC 1.5 – 3.0 dS/m)

d) C-3 Medium salinity waters (EC 3.0 – 5.0 dS/m)

e) C-4 High salinity waters (EC 5.0 – 10 dS/m)

f) C-5 Very high salinity waters (EC> 10 dS/m)

ii) Sodicity (SAR)

1. S-0, Non-sodic waters (SAR <5)
2. S-1, Normal waters (SAR-5-10)
3. S-2, Low sodicity waters (SAR 10-20)
4. S-3, Medium sodicity waters (SAR-20-30)
5. S-4 High sodicity waters (SAR 30-40)
6. S-5 Very high sodicity waters (SAR >40)

iii) Alkalinity (RSC)

1. A-0, Non-alkaline waters (RSC-ve meq/lit)
2. A-1, Normal waters (RSC 0-ve meq/lit)

3. A-2 Low alkalinity waters (RSC tr-2.5 meq/lit)
4. A-3 Medium alkalinity waters (RSC 2.5–5.0meq/lit)
5. A-4 High alkalinity waters (RSC 5.0–10.0 meq/lit)
6. A-5 V. High alkalinity waters (RSC > 10 meq/lit)

Guide lines for Irrigation Water Quality established by the World Food & Agriculture (FAO) Organisation.

Water Constituent	Intensity of Problem		
	No problem	Moderate	Severe
Salinity (dSm^{-1})	< 0.75	0.75-3.0	> 3.0
Permeability (Rate of Infilitration) affected			
Salinity (dSm^{-1})	> 0.5	0.5-2.2	> 2.0
Adjusted SAR: Soils are			
Dominantly montmoinillonite	< 6.0	6-9	> 9.0
Dominantly illite-Vermiculite	< 8.0	8-16	> 16
Dominantly Kalinite -Sesquioxides	< 16	16-24	> 24
Specific ion toxicity Na (as adjusted SAR)	< 3	3-9	>9
Cl (meq l^{-1})	> 4.0	4-10	>10.0
B (meq l^{-1})	> 0.75	0.75-2.0	> 2.0
Miscellaneons			

Water Constituent	Intensity of Problem		
	No problem	Moderate	Severe
NO_3^-–N or NH_4^+–N (meqli⁻)	< 5	5-30	> 30
HCO_3^- (meqli)	<1.5	1.5-8.5	>8.5
pH	6.5-8.4	—	0-5, 9.5 above

Source : Modifed from R.S. Ayres and D.W. West Cot, Water quality for Agriculture, Irrigation and Drainage paper 29, FAO, Rome, 1976.

Chapter EIGHT

Soil and Water Pollution

Agriculture is mainly dependent on the soil and water. The soil covers portion of the surface of the earth. Soil is an intimate mixture of organic and inorganic materials, water and air. Loose powdery soils usually support crop growth. Soil acts as a physical, chemical and biological filter as well as the waste storage for all materials buried and disposed of beneath and on the surface.

Water acts as a universal solvent and cleaner. A good quality water becomes waste water become waste water through the dissolution of different detergents, chemicals like pesticides etc. and other industrial effluents and when this waste water is used for, the irrigation purpose and others then it becomes as serious problem resulting in water pollution.

The Agriculture activities involve addition of nutrients, pesticides and sediments to Soil. On the other hand, industry and urbanization pollute the soil with solid wastes, heavy metals, solvents and several other organic & inorganic substances. Any addition to soil of substances that may exert adverse effects on its functioning can be defined as soil contamination. Since most soils possess buffering capacity, it usually takes sometimes for the negative effects to become visible and then the soil can be considered polluted and the process as soil pollution. Many pollutants reaching the soil can build up to such concentrations which pose serious threats to plants and animal health.

The water is used by various ways like irrigation, livestock, power, industry and domestic consumption. After use from all consumption sources annual return of such waste water is about 286 thousand million cubic meter. Out of this waste water one of the major users is the agricultural sector for irrigating the crops.

There are generally five different kinds of pollution namely (i) Pesticide pollution in soil (ii) Inorganic contaminants or pollutants, mostly heavy metal pollution in the soil (iii) Organic wastes, food processing plants, municipal and industrial wastes dumped on soil (iv) Fertilizer and other salts (v) Radio nuclides.

Pesticides: include insecticides, fungicides, herbicides, rodonticides, nematocides and many other chemical like disinfectants, antibiotics, defoliants, Chemosterilants and juvenile hormones etc. All these chemicals are deadly poisons to plants, animals and human beings. Many toxic pesticides of group organophosphates are active in the environment. When these pesticides (Chemicals) are applied in soils, they undergo following reactions: (i) chemicals may vaporize and be lost to the atmosphere without chemical changes, (ii) may be absorbed by soil colloids, (iii) may move downward through the soil in liquid or solution form and be lost by leaching, (iv) may undergo various chemical reactions within the soil environment or on the soil surface and (v) may be degraded by soil micro-organisms.

A major chemicals like chlorinated hydrocarbons DDT, aldrin, heptachlor etc is their persistence in soil and their movement into water streams through soil erosion and entry in the food chain of various wild life. Their movement in soil, leaching water and absorption by plants and effects on microbiological activity in soils are of great concern.

Some of the pesticides like Heptachlor, Aldrin, Dieldrin are 9 years persistence in soils. Similarly, DDT, BHC and chlordane are 10 years, 11 years and 12 years persistence in soils respectively.

Diuron, Simazine, Atrazine Monuron and 2, 4-D chemicals are persistent in soils 19 months, 17 months, 18 months, 36 months and 30 days respectively.

The hazards due to pesticide pollution, may be somewhat reduced with the application of organic matter in the soil. Some time chemical degradation reduced the intensity of toxicity of pesticides. Bio-degradation by soil micro-organizations is the important method by which pesticides are remove from soils.

Seventy six trace elements all of which are potential pollutants. Number of inorganic contaminants including heavy metals like Hg, Cd, Pb, As, Ni, Cu, Zn, Mn etc. Among these Cd and As are extremely poisonous. Hg, Pb & Ni are moderately poisonous and Cu, Zn and Mn are relatively lower in toxicity.

Sources of Heavy Metal Pollution in the soil-water-air organisms ecosystem:

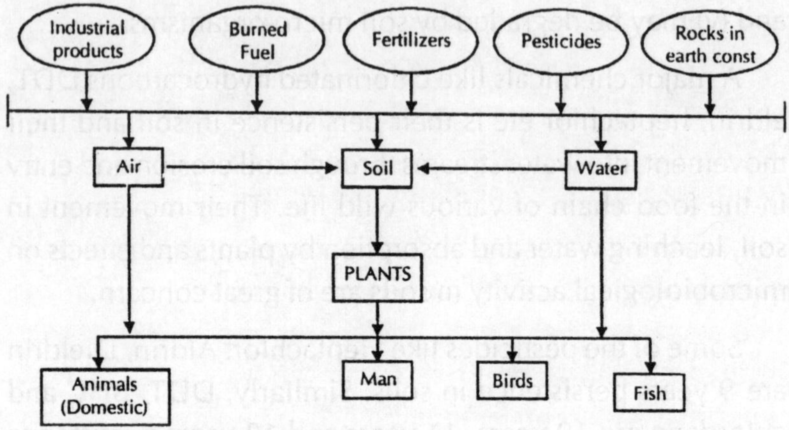

Concentrations (mg Kg^{-1}) of Heavy Metals in Soils & Plants.

Heavy Metals	Soil (mg kg^{-1})	Plants (mg kg^{-1})
Arsenic (As)	0.1-40	0.1-5.0
Cadmium (Cd)	0.1-7	0.2-0.8
Lead (Pb)	2-200	0.1-10
Nickel (Ni)	10-1,000	1

The pollution potential of organic wastes, urban and rural, has become a national and international problem in relation to soil pollution because organic wastes contain detergents, water softners, borates, phosphates, etc. Sewage effluents, agricultural wastes like animal and crop residues etc. cause a serious soil pollution. Sludge contributes soil pollution through the supply of viruses, viable intestinal worm eggs, high concentrations of soluble salts, various heavy metals etc.

Fertilizer and other soluble salts through the application

of chemical fertilizer or through the application of irrigation water threatens production of crops particularly in arid or semi arid regions. The concentrations of fluorine, selenium and boron in water also increases soil alkalinity and create a problem to animals and human beings.

Radio Nuclides are the radio-active materials produced from explosions and the dumping of the waste materials from the nuclear power plants, nuclear reactors and waste from the medical and other research laboratories. The most important and long lived radio-active elements in soils are strontium (-90) [half life = 28 years] and Cesium (-137) [half life= 30 years]. Radio active elements could be decontaminated by continuous cropping, removal of surface soil, deep ploughing, leaching, use of fertilizers and amendments. The use of chelates like EDTA, DTPA in conjunction with algal and fungal cells is most effective in reducing radio activity from the soil.

Gaseous emissions from industrial areas, burning fossil

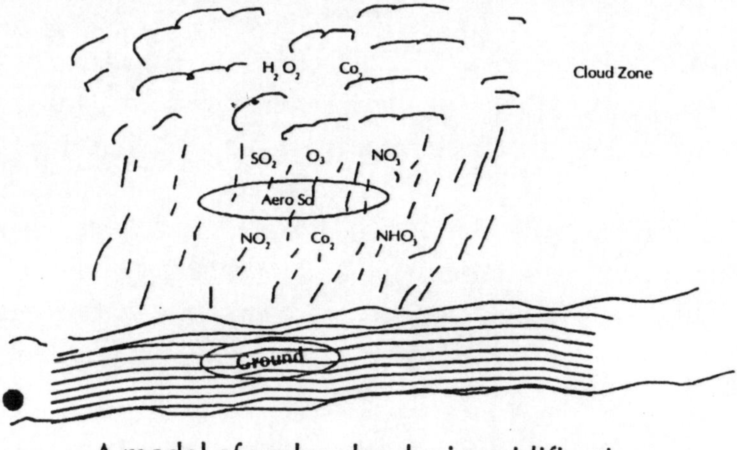

A model of under cloud rain acidification

fuels, automobiles, smelting and mining etc. very frequently leads to soil and water pollution through entry into terrestrial ecosystem as acid rain or directly as particulate matters on the soil surface. Reactions involved in the acid rain :

$$2\,NO + O_2 \rightarrow 2NO_2 + H_2O \rightarrow HNO_3 + HNO_2$$

$$2\,SO_2 + O_2 \rightarrow 2SO_3 + 2SO_3 + 2H_2O \rightarrow 2H_2SO_4$$

Index